GW00838407

Minerals of Cornwall and Devon

Minerals of Cornwall and Devon

P. G. Embrey
and R. F. Symes

British Museum (Natural History), London
and the Mineralogical Record Inc.,
Tucson, Arizona

ACKNOWLEDGEMENTS

We are indebted to many friends and colleagues for their generous contributions to and patient help in the production of this book.

In the British Museum (Natural History), Frank Greenaway and Colin Keates have taken nearly all of the photographs. A.G. (Coup) Couper and John Fuller, the latter now sadly deceased, spent much time in selecting the mineral specimens. Valerie Jones and Alan Hart drew the maps and other line illustrations. Chris Stanley supplied many valuable comments, and Chris Owen, as well as being the editor, selected several of the older mining pictures from the collections of the County Museum, Truro.

In supplying material from the library and collections in their care, and the answers to innumerable questions, the curators of the County Museum, Leslie Douch and Roger Penhallurick, have been unfailingly friendly and generous with their time.

Wendell Wilson, the editor of the *Mineralogical Record* – the American co-publisher of this book – offered much guidance during the preparation of the text. Chris Halls and Eileen Brunton generously supplied the 'disc' of their own bibliography, to which we had contributed, to help in its production. Ivor Cornish, of Ambra Books, Bristol, and Mike Richards, of the Truro Bookshop, have enabled one of us (PGE) to buy many of the books listed.

In extending our special thanks also to Kenneth Hosking, Martin Mount, Courtney Smale, George Ryback, Bryan Lloyd, Ron Cleevely, Tom Vallance, Graham Durant, and Robert Jones, we hope that we give no offence by inadvertent omissions.

This list would not be complete without acknowledgement of the great debt of gratitude owed to our late friends Sir Arthur Russell, Arthur Kingsbury, and Richard Barstow.

PGE RFS

First printed 1987
British Museum (Natural History)
Cromwell Road
London SW7 5BD

Embrey, P. G.

Minerals of Cornwall and Devon.
1. Mines and mineral resources——England
——Devon 2. Mines and mineral resources
——England——Cornwall
I. Title II. Symes, R. F.
549.9423'5 QE262.D45
ISBN 0-565-00989-3 paperback
ISBN 0-565 01046 8 hardback

Printed in England by Balding + Mansell UK Limited

CONTENTS

INTRODUCTION

The need for a well-illustrated account of the minerals of Cornwall and Devon has, for long, been apparent to mineralogists and collectors of all persuasions. In this book Peter Embrey and Robert Symes have made an original and idiosyncratic contribution, reflecting a long-held fascination with historical and scientific aspects of the subject. Such a work could easily be justified on aesthetic as well as on scientific grounds, but the fact that mercantile, industrial, and scientific development in Britain has been so intimately related to the exploitation of the mineral fields of the south-west peninsula lends additional historical, social and economic significance to the theme.

The richness and variety of mineral species found in the great network of lodes woven through the granite and killas of the peninsula has, since antiquity, supplied metals for trade and industry. Later, the geological phenomena of the region attracted the attention of scientists and collectors through whose work fundamental contributions were made, to the knowledge of mineral chemistry, morphology and paragenesis. The difficulties confronted in mining the tin, copper and other ores from their hydrothermal repositories deep underground, of necessity encouraged the development of special skills and the invention of machines by which the workings could be made safer and more profitable. These innovations have been carried to other mining fields throughout the world by pioneering Cornishmen and the language of mining and mineralogy itself has been enriched with many words of Cornish origin which continue in daily use. Thus it is that a wealth of literature concerned with the topography, geology and mineralogy of the region is complemented by an equally fascinating record of mining, engineering and metallurgy in which the names of Newcomen, Trevithick, Woolf and Bickford take their place beside those of Pryce, Klaproth, Raspe, De la Beche, Henwood, Rashleigh and many others.

By the inclusion of chapters on the geology of south-west England and the history of the mines and mining, together with an account of some of the collectors and dealers to whom the minerals of the region gave scientific inspiration, aesthetic satisfaction and financial reward, the authors have opened this treasury of historical and scientific references which the specialist and amateur alike will value. For illustration some eighty exemplary and beautiful specimens have been chosen, mainly from the collections of the Department of Mineralogy of the British Museum (Natural History) where the greater part of the mineralogical inheritance of this famous region has been preserved, amplified and enriched by the recent incorporation of the collections of the Geological Museum. These collections held by the Museum in association with a unique library of books and papers concerning the geology and mineralogy of the mining fields of Cornwall and Devon form a resource appreciated internationally. Through the publication of *Minerals of Cornwall and Devon*, Peter Embrey and Robert Symes, assisted by their colleagues in the British Museum (Natural History), have made that precious resource more accessible to us all.

Christopher Halls
Royal School of Mines
August, 1987

Fig. 1 Thin section showing zoned tourmalines with feldspar and quartz from the 'tourmaline floor' at Grylls Bunny, Botallack (see p. 36, and Jackson, 1974). Photo C. Halls.

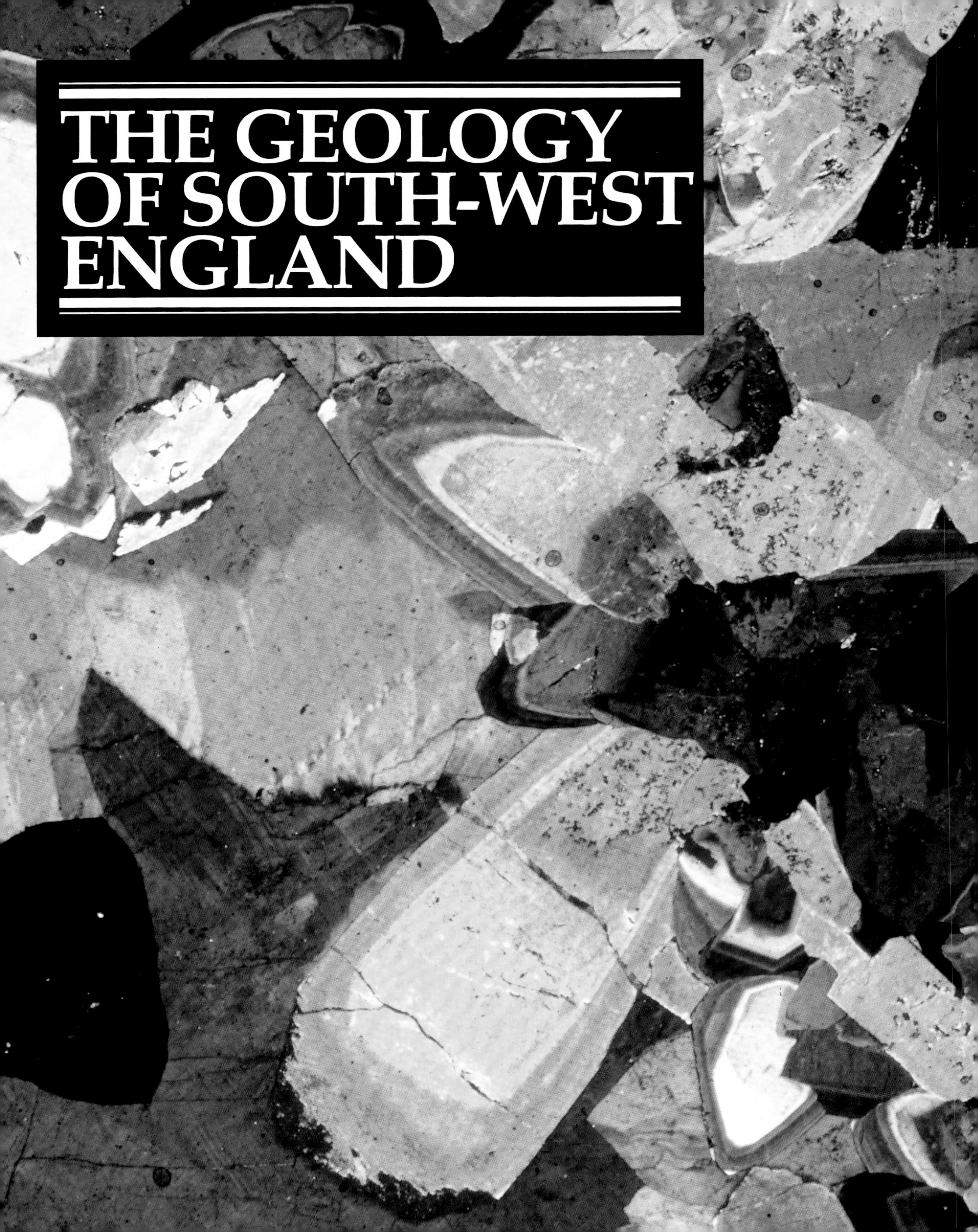
THE GEOLOGY
OF SOUTH-WEST
ENGLAND

Introduction

The metallogenic province of south-west England is a region about 100 miles long, and comprises the whole of Cornwall and much of western Devon, including the Dartmoor granite. Cornwall has long been a classic area of British Geology, on account of its wealth of mineral deposits and the variety of its igneous and metamorphic rocks. Devon is also rich in rock types, and is known to geologists world-wide as the type area of the Devonian System, established in 1839 by Sedgwick and Murchison (fig.2).

So much has been written about the general and detailed geology of the region that yet another review cannot be entirely original. All references that we have consulted appear in the bibliography, but we would recommend to the general reader the following treatments of the subject: *British regional geology – south-west England* (British Geological Survey, 4th ed.); *Geology of Devon* (eds Durrance & Laming), 1984; and Thorne & Edwards, 1985.

The region is peninsular, with no point much further than 25 miles from the sea, and has strong maritime traditions. It is also well known for its contrasting scenery. The rugged cliffs of the north and west coasts, facing the Bristol Channel and Atlantic Ocean, provide some of the most magnificent coastal scenery in the British Isles. Much of the southern aspect is more gentle, with hills rolling down to the English Channel, but in places cliff scenery on the south coast can be both spectacular and colourful. Headlands of more resistant rocks project from a coast indented by drowned river mouths, notably those of the Tamar at Plymouth and the Fal at Falmouth (fig.3). East Devon, with its hills and valleys of Cretaceous rocks, and the red soils derived from the rocks of the Permian and Triassic, has a character that contrasts with the Devonian and Carboniferous pasture land of mid-Devon and Cornwall and with the Devonian sandstone moorland of Exmoor further north.

The topography of the peninsula is dominated by rounded masses of granite, topped by characteristic rugged grey tors and surrounding rock debris, standing out above moorland or rough pasture (fig.4). Early travellers described the region as mountainous, a considerable overstatement. The tops of the granite masses get progressively lower towards the west. Dartmoor is the highest, averaging some 1400 ft above sea level, with a maximum of 2039 ft at High Willhays. Brown Willy, at 1375 ft on Bodmin Moor, is the highest hill in Cornwall; and the high point of the Isles of Scilly, the most westerly part, is a mere 166 ft. Often the land-

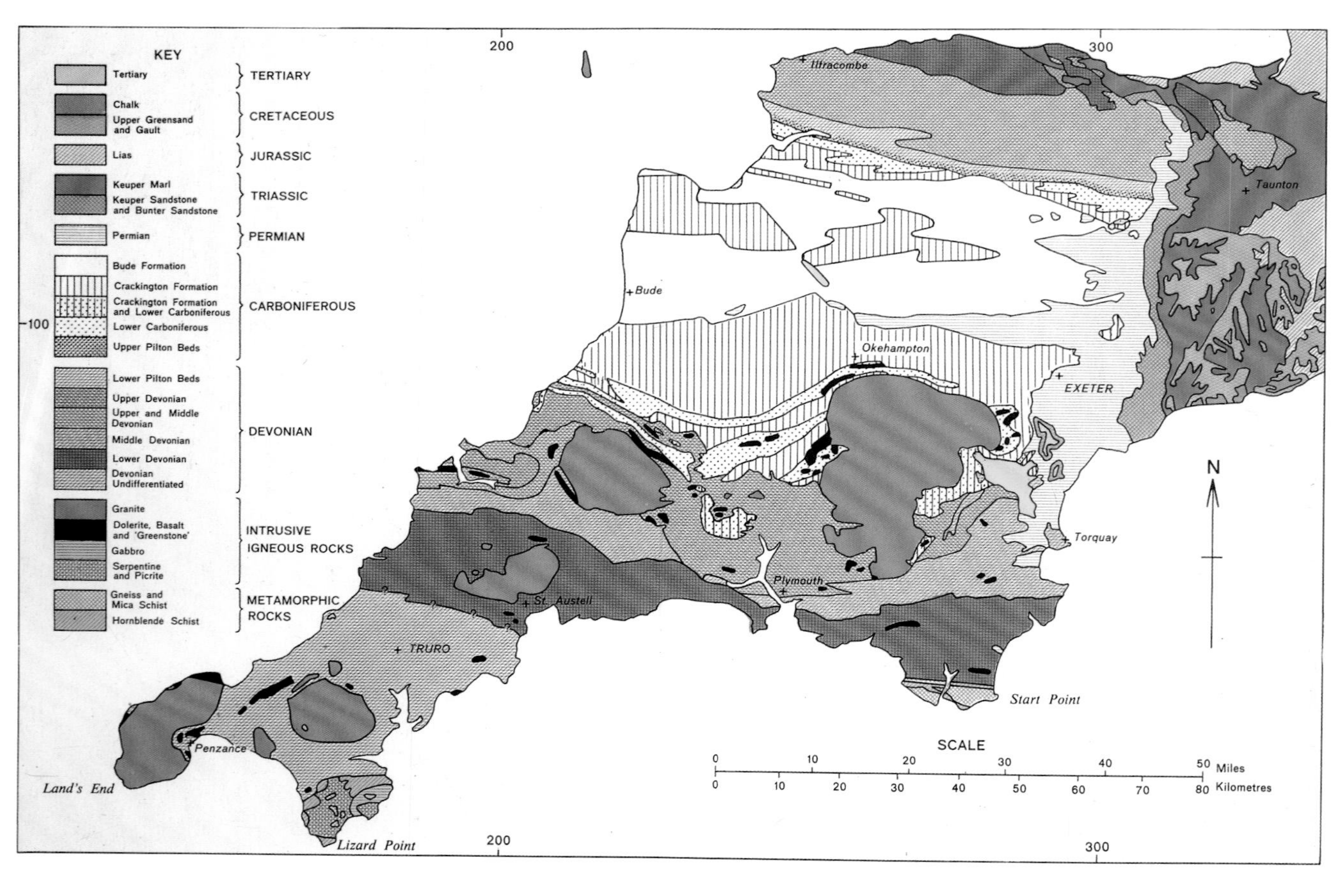

Fig. 2 Geological map of south-west England (reproduced by permission of BGS).

Fig. 3 Falmouth, Cornwall from Pendennis Castle (T. Allom, 1831).

Fig. 4 Haytor Rocks, East Dartmoor, Devon. A typical tor formed from coarse biotite granite.

Fig. 5 Country view close to St Agnes, north Cornwall includes the remains of the Wheal Charlotte engine house. In south-west England farming and mining existed side by side for many centuries.

scape is relatively bare and treeless, with solitary engine houses and scarred ground to remind us of past mining activity (fig. 5).

Geological history

Most of the region consists of strongly-deformed sediments, which were intruded by granites in late Carboniferous or early Permian times, about 300 million years ago. The granites, and the deformation of the sedimentary rocks, are manifestations of the Variscan orogeny. At the end of the Carboniferous period, and in early Permian times, before the present Atlantic Ocean had been developed, the margins of North America and Europe lay fairly close together. Within this framework (fig. 6), the Variscan orogen was a broad belt of deformation running from what is now central Europe to the east coast of North America, including the southern British Isles and fringing the north-west coast of North Africa.

There is no generally-agreed and clear picture of the plate-tectonic setting at this time. It seems likely, however, that two continental masses, including what are now southern and northern Europe, were moving together; that there was subduction of oceanic crust between the two; and that Cornwall (Cornubia), with the rest of south-west England, lay on the southern edge of the northern block. Several models have been proposed; see, for example, Bromley (1976), Badham (1982), and Floyd *et al.* (1983).

The slaty shales and mudstones, with subordinate bands of sandstones and conglomerate, which form by far the greater part of the region, have long been known collectively by the local term 'killas'. These sediments are of Devonian and Carboniferous age, and both successions of rocks contain contemporaneous beds of lava and tuffs (see fig. 7), and basic igneous intrusives (known locally as greenstones). Most of Cornwall is composed of Devonian rocks of this type, in which recognisable fossils are sparse (but see House, 1965); whereas the Devonian rocks of north Devon, although varied, consist mainly of coarser sandstone lithologies. In south Devon, associated with slates, there are some thick, often reef, limestones. Carboniferous rocks form the central part of Devon, and a little of the north Cornish coast.

Due to the earlier Caledonian earth movements, by the start of the Devonian (about 400 million years ago) most of Britain formed part of the 'Old Red Continent'. Mountainous regions lay to the north, and to the south a coastal plain bordered open sea. Sediments from this landmass drained southwards, through deltas, to an elongate subsiding marine trough covering most of what is now Devon and Cornwall.

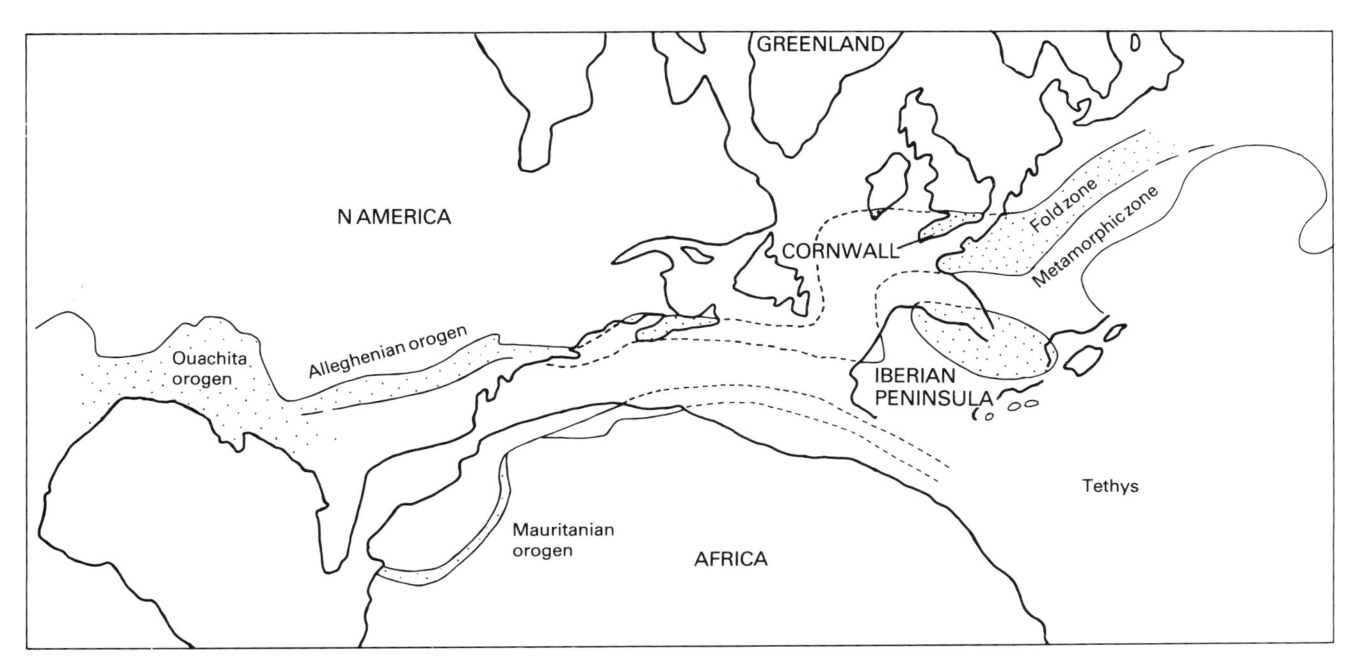

Fig. 6 Sketch map of the extent of the Hercynian (Variscan) orogenic belt on a pre-Mesozoic reconstruction of the north Atlantic area (after Riding, 1974).

Various sediments accumulated in these Devonian coastal plains, deltas, and open sea, getting finer-grained as the northern landmass was eroded. The water was generally deeper over what is now south Devon and east Cornwall, leading to the formation of thick muds, but local areas of clear shallow water permitted the growth of the coral reefs which can now be seen in the limestone cliffs of Plymouth and Torquay. In west Cornwall, dark muds and coarse sands (turbidites) were accumulating at the same time, derived mainly from a southerly source. Volcanic activity was producing submarine lavas over a wide area, such as the pillow lavas of Pentire Point, on the north Cornish coast.

In most areas, this general pattern of sedimentation continued beyond the Devonian. During the Lower Carboniferous, the dominant sediments were mudstones, some of them calcareous, with volcanic activity represented by lavas, tuffs, and agglomerates, such as those at Brent Tor. Thick beds of cherts and dark limestones and, as the sea became shallow, river-derived sands and muds with carbonaceous matter ('culm'), accumulated over a wide area. In the Crackington Formation of the Upper Carboniferous, there is a sequence of shales with turbidite sandstones; and in the overlying Bude formation, thick-bedded and massive sandstones are interbedded with siltstones and shales.

Fig. 7 Spilitic pillow lavas (Upper Devonian) at Chipley quarry, south Devon.

The Devonian and Carboniferous sediments were then folded and faulted during the Variscan orogeny, a major earth-movement event spanning late Devonian to late Carboniferous times. Interpretations of the structure of the region indicate a number of tectonic zones, distinguished by fold style. The deformation occurred in several phases, producing a series of complex major folds with a general east-to-west trend, and northerly-translated thrusts; but local deformation may modify this general pattern. In north and central Devon the structures form an apparent synclinorium, of Culm facies rocks of Carboniferous age, largely devoid of mineralization. In north Cornwall the structures are complex, chevron and other fold types being well displayed on the coast (fig.8). Large areas of central Devon and Cornwall consist of thin-skinned thrust and nappe structures.

The rocks of the Lizard area, Cornwall, are believed to be the oldest in the region (fig.2). The Lizard Complex is an assemblage of peridotites, gabbros, dolerites, amphibolites, acid and basic gneisses (Kennack Gneiss), and metamorphosed sediments. Most of the ultrabasic rocks have been serpentinized. Such recent studies of the Lizard complex as those of Styles (1982) have suggested that it

Fig. 8 Cliff section at Millook Haven, north Cornwall, showing zig-zag folding and overthrusting in Upper Culm shales and mudstones (Crackington Formation).

consists of tectonic units, separated by low-angle thrusts, the upper unit showing a distinctive ophiolite succession. Underlying these tectonic units, the gneisses may have been derived from a series of sediments and lavas. The Lizard complex could, therefore, have originated as a fragment of ocean floor, of Lower Palaeozoic or earlier age, that was thrust upwards (obducted) during the Variscan orogeny.

The Variscan orogeny produced only a low-grade metamorphism in the Devonian and Carboniferous rocks. In south Devon, however, the spectacular headlands of Start Point and Bolt Tail are composed of schists, separated from the low-grade slates to the north by a major tectonic feature.

The overall effect of the earth movements was to produce folding, jointing, and cleavage in these earlier rocks, about an axis with a roughly ENE-WSW trend. This deformation episode appears to have been largely completed before the emplacement of the Cornubian granite batholith, during late-Carboniferous and early-Permian times, and the thermal metamorphism associated with it. Fissures, probably related to the consolidation and cooling of the granite, opened along the same general trend, and some of these filled with magma to form the quartz-porphyry dykes known as 'elvans' (see p.12). Episodes of fracture-filling, or 'lode' mineralization, caused by hydrothermal mechanisms, followed: as a general rule, tin and copper minerals are found in roughly E–W lodes, and later lodes of roughly N–S trend carry lead, zinc, and iron.

Following these Variscan magmatic and orogenic events, sedimentation recommenced in Devon under semi-arid conditions, with coarse clastic deposits forming in intermontane valleys and low-lying areas. This type of sedimentation continued until the end of the Triassic, about 180 million years ago, when marine conditions were re-established leading to subsequent shelf-sea sedimentation of the east Devon Mesozoic rocks.

During the late Palaeozoic, and perhaps the early Mesozoic, considerable uplift occurred. A mountain chain emerged, into a semi-arid climate which allowed deep weathering and erosion to expose the tops of the granite domes.

The Variscan structures of Devon are cut by NW-SE wrench faults (Dearman, 1963), which considerably influence the structural pattern of the region. The Sticklepath fault (north-east Dartmoor) is such a major dislocation, and had considerable influence on the formation of the Bovey Tracey basin, a site for the valuable Tertiary ball clay deposit. In the Cenozoic period, weathering continued in a sub-tropical climate, with uplift consequent on Alpine earth movements.

Periglacial conditions prevailed during much of the Quaternary period, within the last 3 million years; wave-cut platforms and raised beaches are associated with this period. The last glacial stage ended some 10 000 years ago, and the warmer climate completed the fashioning of scenery, including the drowning of river mouths as the sea level rose.

Recent studies of stanniferous placer deposits (Camm & Hosking 1984,1985) show that their genesis is complex, and that most of them formed by liberation of cassiterite from associated minerals with reworking during the Mio-Pliocene, followed by further transport and deposition during the Pleistocene to Recent. The first tin recovery in the region was from these deposits.

The Granites

The Cornubian batholith is spatially associated with the base metal mineralization, and is composed of several chemically similar but distinct intrusions. It extends for 100 miles, with a general NE–SW strike, and forms the spine of the peninsula. The granite is expressed at surface in a series of separate plutons: Dartmoor (the largest), Bodmin, St Austell, Carnmenellis, Tregonning–Godolphin, and Land's End, with a few smaller related masses such as Carn Brea, Carn Marth, and St Agnes Beacon. Further westwards it forms the Isles of Scilly. Geophysical

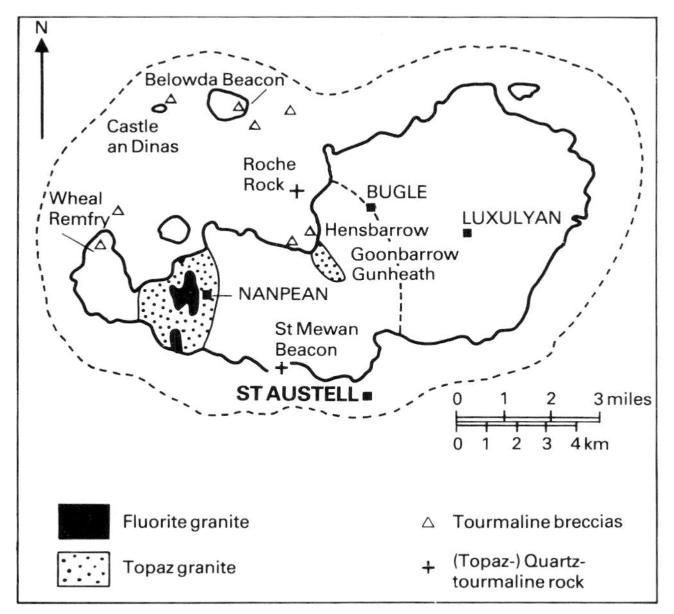

Fig. 9 Sketch of the geology of the St Austell granite, showing the outcrops of topaz and fluorite granite. Biotite granites (blank) show considerable variation in texture; in the western half strongly kaolinised zones are marked by china clay pits. The extent of the contact metamorphic aureole is marked by the dotted line (after Manning, 1985).

determinations have shown that the exposed plutons are the surface expression of a continuous intrusive body.

The granite, more strictly an adamellite, has been subdivided by many workers into a number of petrographically distinct types. By far the dominant of these, and the earliest from the evidence at the present level of erosion, is a medium- to coarse-grained biotite granite; the size, amount, and alignment of the typical K-feldspar megacrysts are variable. This biotite-granite includes xenoliths of altered igneous and sedimentary rocks. The main masses are accompanied by a wide variety of minor intrusive granitic rocks, including porphyry dykes, pegmatite and aplite dykes and veins, and hydrothermal breccias.

Other important types of granite constitute a separate chemical and mineralogical grouping characterized by the presence of such minerals as Li-bearing micas, albite, topaz and fluorite. Some authors have suggested that these, and other variations in granite petrology, have been produced by metasomatism of pre-existing granite types, a good example being parts of the St Austell pluton (fig.9). The development of late magmatic fluids gave rise not only to extensive metasomatism, resulting in the growth or enlargement of potash feldspar megacrysts (as in the roof-zone of the Land's End pluton) and some pegmatites, but also to some metalliferous mineralization.

The intrusion of the batholith occurred in several pulses, of similar magma types, over a considerable period. This enabled local production of derivative magmas, by fractional crystallization or remelting. The granite was emplaced by both forcible intrusion and assimilation of the country rock (killas) through stoping. Some authors have suggested that the limited width of the contact thermal aureole around the intrusion indicates that the granite was emplaced in a partially-solidified state rather than as a fully-molten magma.

Contact metamorphism

The granites baked the surrounding country rocks, normally producing 'spotted slates', but the effects vary with the nature of the host rocks. The argillaceous Devonian strata around the Land's End granite, for example, were converted to cordierite-bearing hornfelses; whereas, in the Carboniferous argillaceous rocks north of Dartmoor, andalusite takes the place of cordierite. In calcareous rocks, such as the limestones associated with cherts and calcareous shales, along the northern margin of Dartmoor in the Meldon area, hornfelses containing pyroxene, garnet (grossular), and wollastonite occur; and there was local formation of scapolite, idocrase, bustamite, and rhodonite.

Kaolinization

All the exposed granites of the peninsula exhibit some degree of kaolinization. China clay has been extracted at various times, and for several years has been the most important mineral exported from the British Isles. At present, the chief workings are on the western part of the St Austell outcrop (fig.10), with others on the south-west side of Dartmoor (Lee Moor) and on Bodmin Moor to the north and west. The changes involve the conversion of plagioclase in the granite to secondary mica and kaolinite-mica aggregates, and some of the K-feldspar to secondary mica. The biotite is often altered to a non-hydrous secondary mica. Crystals of plagioclase feldspar tend to undergo the greatest alteration, but K-feldspars were usually more resistant and only partially altered; at

Fig. 10 General view overlooking Goonbarrow China Clay Pit, St Austell, Cornwall. Within this same area Wheal Martyn, nearby, is now the site of the China Clay Museum with comprehensive exhibits of the industry (photo C. Halls).

a few localities these have been completely converted to masses of sericitic mica, whilst retaining their normal twinned form, and may be seen in many mineral collections bearing the clay-workers' name 'pig's eggs'.

The degree of alteration of the granite increases towards fissures such as joints or faults. Theories of the probable mechanism involved in kaolinization change with bewildering frequency, one current belief being that the alteration was caused by circulating acid, aqueous fluids, but the exact age and nature of these fluids, whether magmatic or meteoric or both, is still the subject of active research.

A related, but distinct type of granite alteration is the local formation of 'chinastone', mainly in the St Austell area, where partial kaolinization is accompanied by enrichment in fluorite.

Mineralization

Introduction

The main emphasis of this publication is on the exotic and fine crystallized specimen mineralogy of Cornwall and Devon, rather than being a detailed survey of ore genesis. It is, nevertheless, the primary mineralization, especially of the base metals, and subsequent alteration and redistribution that has provided so many fine mineral specimens, whether they be considered as ore or gangue. Some of the main ore and gangue minerals are shown in fig.11, and the nature and origin of the mineralization calls for a brief description.

From the mineral collector's point of view, one of the essential features of the mineralizing process is that, during the formation of the deposits, conditions should have been such that sufficient space (within vughs) was available for the free growth of crystals, and that subsequent solution or tectonic events acted to protect and even enhance crystal groups rather than destroy them.

Nature and origin

Pegmatites and aplites were formed, in the roof zones of the batholith and associated bosses, at a late stage of the emplacement of the granites (fig.12). Radiometric dating suggests that this phase was largely complete 290 million years ago.

Fig. 11 Generalised paragenetic sequence of ore and gangue minerals in south-west England (after Stone & Exley, 1985).

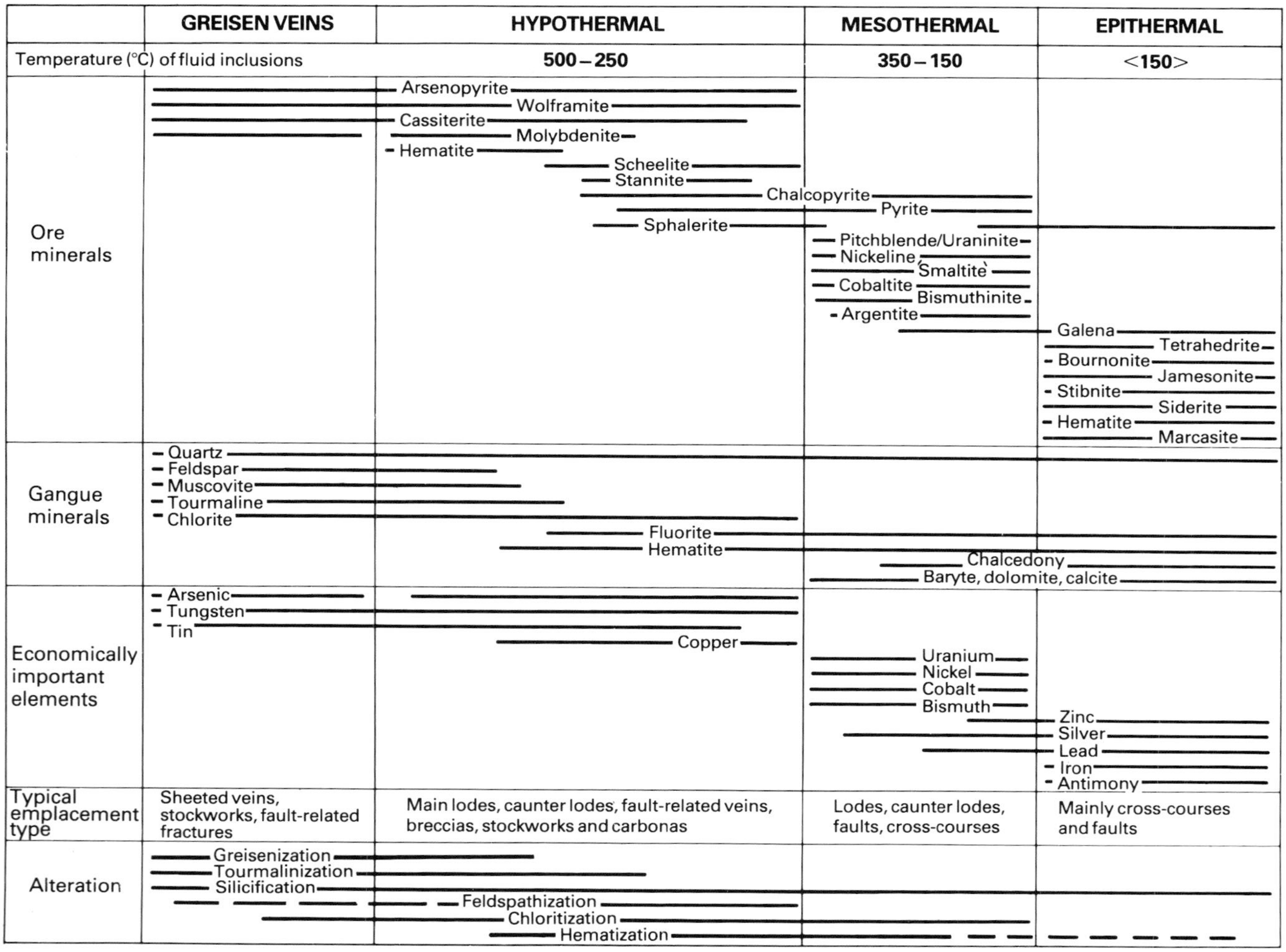

	GREISEN VEINS	HYPOTHERMAL	MESOTHERMAL	EPITHERMAL
Temperature (°C) of fluid inclusions		500 – 250	350 – 150	<150>
Ore minerals		Arsenopyrite, Wolframite, Cassiterite, Molybdenite, Hematite, Scheelite, Stannite, Sphalerite	Chalcopyrite, Pyrite, Pitchblende/Uraninite, Nickeline, Smaltite, Cobaltite, Bismuthinite, Argentite	Galena, Tetrahedrite, Bournonite, Jamesonite, Stibnite, Siderite, Hematite, Marcasite
Gangue minerals	Quartz, Feldspar, Muscovite, Tourmaline, Chlorite	Fluorite, Hematite	Chalcedony, Baryte, dolomite, calcite	
Economically important elements	Arsenic, Tungsten, Tin	Copper	Uranium, Nickel, Cobalt, Bismuth	Zinc, Silver, Lead, Iron, Antimony
Typical emplacement type	Sheeted veins, stockworks, fault-related fractures	Main lodes, caunter lodes, fault-related veins, breccias, stockworks and carbonas	Lodes, caunter lodes, faults, cross-courses	Mainly cross-courses and faults
Alteration	Greisenization, Tourmalinization, Silicification	Feldspathization, Chloritization, Hematization		

Fig. 12 Roof zone of the granite cupola and killas country rock at Porthmeor Cove, Penwith, west Cornwall. Differentiation of the granite in situ has led to the formation of aplitic and pegmatitic layers, comb layering of potash feldspars and pockets of tourmaline.

After this, but before the main phase of mineralization, during the development of fissuring, intrusive tourmaline-rich breccias were formed (such as the 'explosion' breccia at Wheal Remfry, near St Austell) (fig.13). The possible mechanism for this was the release of gaseous components from the granite through the initial fractures, which may also have tapped residual magma to form the characteristic porphyritic granite dykes known as 'elvans'. In some areas, tin– tungsten mineralization is associated with this magmatic activity.

The next stage of the consolidation and cooling process was the evolution of hydrothermal fluids, leading eventually to the major stage of mineralization in which the deposition of predominantly tin-tungsten-copper minerals in joints and fissures produced the main economic concentrations.

Fig. 13 Intrusive hydrothermal breccia at Wheal Remfry, St Austell, Cornwall showing clasts with original tourmaline layering (photo C. Halls).

Various types of mineralized structure are recognized. The following are those that have usually been distinguished, but it needs to be emphasized that they are rarely as clean-cut as the list suggests; for example, the wall rocks of a lode may well contain economic amounts of tin and show extensive alteration.

a) **Mineralized faults, or lodes.** In south-west England, faulting has played a major part in governing the pattern of lode development, and veins are often (but not always) directly related to normal faults. A general fracture pattern, which eventually provided channels for mineralizing fluids, probably developed during folding of the country rocks and intrusion and consolidation of the granite (fig.14).

Only a minority of all fissures have been mineralized. Those that have been, the lodes, are not uniformly rich in metallic ores and their content varies both laterally and in depth. Richness is often greatest at intersections with other structures, but shows no clear relation to the host rock; in two neighbouring mines, the lode was rich within the granite and poor in the killas at Great Work mine, and the reverse held true at Wheal Vor (Hill & MacAlister, 1906:172).

Within each mining district, the principal vein systems are grouped into:

1) Normal lodes, which follow the mean structure direction of local rocks;
2) 'Counter' (or 'Caunter') lodes, which intersect normal lodes at a small angle; and
3) 'Cross-courses', which are roughly at right angles to normal lodes, and often occur as NW–SE trending mineralized faults displacing tin- or copper-bearing lodes.

The lodes are generally narrow structures, but can vary considerably in width. They are commonly near-vertical or steeply dipping, although some economically important lodes have a much gentler dip, for example the Great Flat lode of the Camborne-Redruth area, has an average 40° dip (fig.17). In general, the regional strike of the lodes is ENE–WSW, tending to E-W in east Cornwall and Devon (fig.18); but these trends have been influenced locally by tectonic and intrusive patterns.

They may be simple infilled structures, with mineral banding parallel to the walls, or composite due to successive periods of mineralization (fig.14). Later relaxation of the regional stress, and the cooling of the granite, caused several phases of reactivation of these fractures, often with the sequential deposition of lower-temperature ores in the re-opened lodes. Evidence of reactivation is provided by repetitive banding, by mineralized fault breccias, and sometimes by shattered and altered wall rocks

b) **Replacement structures.** In places, chloritization and silicification, often accompanied by tourmaliniza-

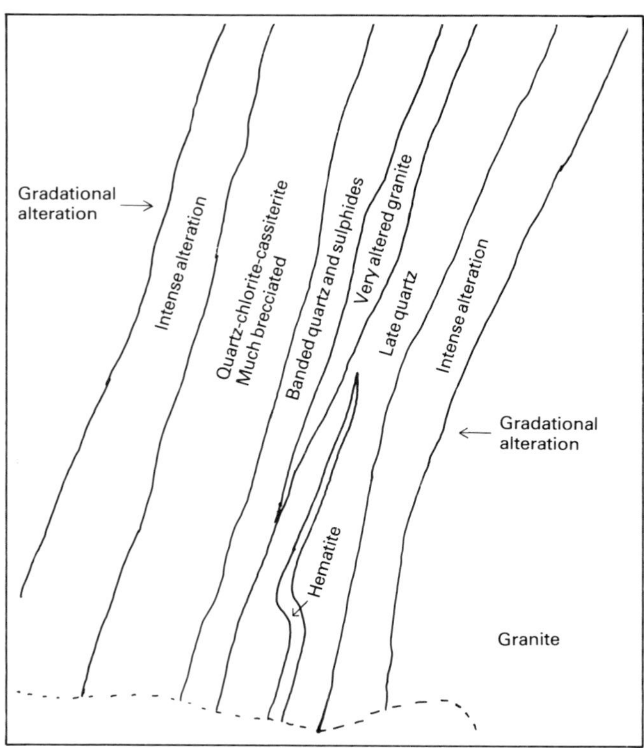

Fig. 14 Face of 14 No. 2 Branch Lode, Geevor Mine. A typical example of fissure-vein type mineralization. The lode comprises composite quartz veining indicating successive reactivation of a single fissure cutting coarse-grained megacrystic granite. Three main events are illustrated: left-hand side comprising early quartz-chlorite-cassiterite; central, later quartz sulphide veining (mainly copper sulphides); and right-hand side a late barren quartz event with traces of hematite. The adjacent wall rock alteration increases in intensity with proximity to the veining. Alteration with early tourmalinization, followed by sericitization and chloritization and later hematization. Vein approximately 1 metre wide (photo M. Mount).

tion, lead to replacement of granite and country rock. Often, there is no obvious fissuring structure to the rocks. Where the metasomatic fluids were rich in metallic elements, then pockets of mineralization formed. Wall-rock alteration is common adjacent to the lodes of the south-west, but rarely extends for more than a few yards, and is often unsymmetrical between hanging wall and footwall. Bands of 'flucan', or clay, parallel to the lode or cross-course, are common. Various names have been given by the miners to the different types of alteration: 'caple' was hard and black, due to quartz-tourmaline alteration; and 'peach' was generally green and much softer, due to chloritization. Hematization is common.

Carbonas, pipes, and floors are names given to rich, localized replacement, apparently formed by solutions migrating along minute fractures. In the case of floors, these sub-horizontal ore bodies, usually associated with tourmaline, were produced by mineralizing fluids in flat-lying joints or faults.

c) **Stockworks.** These are bodies of rock in which the mineralization is contained in numerous veinlets, of varying strike and dip, which form a netted or interlacing mesh. Usually one direction of veinlet is dominant and richer in ore. Usually ore grades are low, and deposits are worked by open pit methods. By far the largest and most-explored stockwork deposit in south-west England is the cassiterite–wolframite–arsenopyrite deposit at Hemerdon, north-east of Plymouth, which is confined to the northern part of the Hemerdon Ball granite. Within this stockwork, the granite is more-or-less kaolinized.

d) **Stratiform deposits.** These occur in a few areas, one of the best examples being the bedded copper deposits at Belstone and Ramsley, on the northern margin of the Dartmoor granite. Further south, at Haytor mine, magnetite was worked in stratiform deposits from thermally metamorphosed sandstones and shales.

Within these structures, various associations of the tin-tungsten-copper mineralization can be recognized:

1) Early metasomatic changes often occurred both in the granite and in the surrounding country rocks. Good examples of the latter occur in the greenstones of the Botallack–Levant area, St Just, where interbedded shales, pyroclastics, and lavas have been extensively replaced by garnet–magnetite–axinite skarn assemblages containing tin-bearing minerals.

2) Formation of greisen-bordered vein structures, one of the best known examples being that at Cligga Head, Cornwall (figs.15 and 16). Many of the veins are spaced so closely that pervasive alteration has left no fresh granite. Other examples of considerable economic

Fig. 15 Cligga Head quarry, Perranporth, Cornwall, showing multiple greisen veins (photo C. Halls).

interest are to be found at Hemerdon, Devon, and Redmoor in east Cornwall.

3) Main, or lode-type base-metal mineralization, in which the major economic tin–tungsten and copper ores occur.

Other metallic associations have been mined at one time or another. Arsenic was recovered as a by-product of the roasting and smelting of tin–tungsten–copper ores, in which it occurred as arsenopyrite, sometimes as löllingite. Gold was not uncommon in the gossans above ores, but was more abundant as grains and small nuggets in the tin placers. Silver occurs, sometimes in

Fig. 16 Detail of greisen lode at foot of cliff section, Cligga Head showing a thin mineralized vein and greisen borders.

significant quantities, in the Pb–Zn veins of Devon and east Cornwall, and was also found in association with Ni–Co–Bi–As–U ores in cross-courses close to altered greenstones. Antimony mineralization, generally recognized as 'stratabound', has a close spatial relationship with spilites and tuffs in north Cornwall and south Devon. Iron and manganese oxide ores, the latter most common in the chert beds of the Culm, were of local importance.

Phases of deuteric activity were widespread in the region, causing sericitization, tourmalinization, and greisening; the latter being essentially the replacement, in varying degrees, of perthite in the granite by quartz and secondary mica, often with topaz. Tourmaline and quartz formation took place at all stages of cassiterite mineralization. Alteration due to tourmalinization can produce rocks containing little but quartz and tourmaline, such as the quartz–schorl rock of Roche Rock, near St Austell; and the decorative luxullianite, from Luxulyan, is an intergrowth of red and pink feldspar with acicular blue-black tourmaline.

Supergene depletion and enrichment

Secondary alteration, particularly of the base metal sulphides above the water table, seems complete throughout the region; with few exceptions, however, these zones of leaching and enrichment (alternatively, of oxidation and reduction) have been removed by erosion or worked away by man. The changes took place by the action of downward-percolating surface waters, as indicated in fig.19. The upper parts of the lodes are commonly iron, less frequently manganese, oxide 'gossans', together with

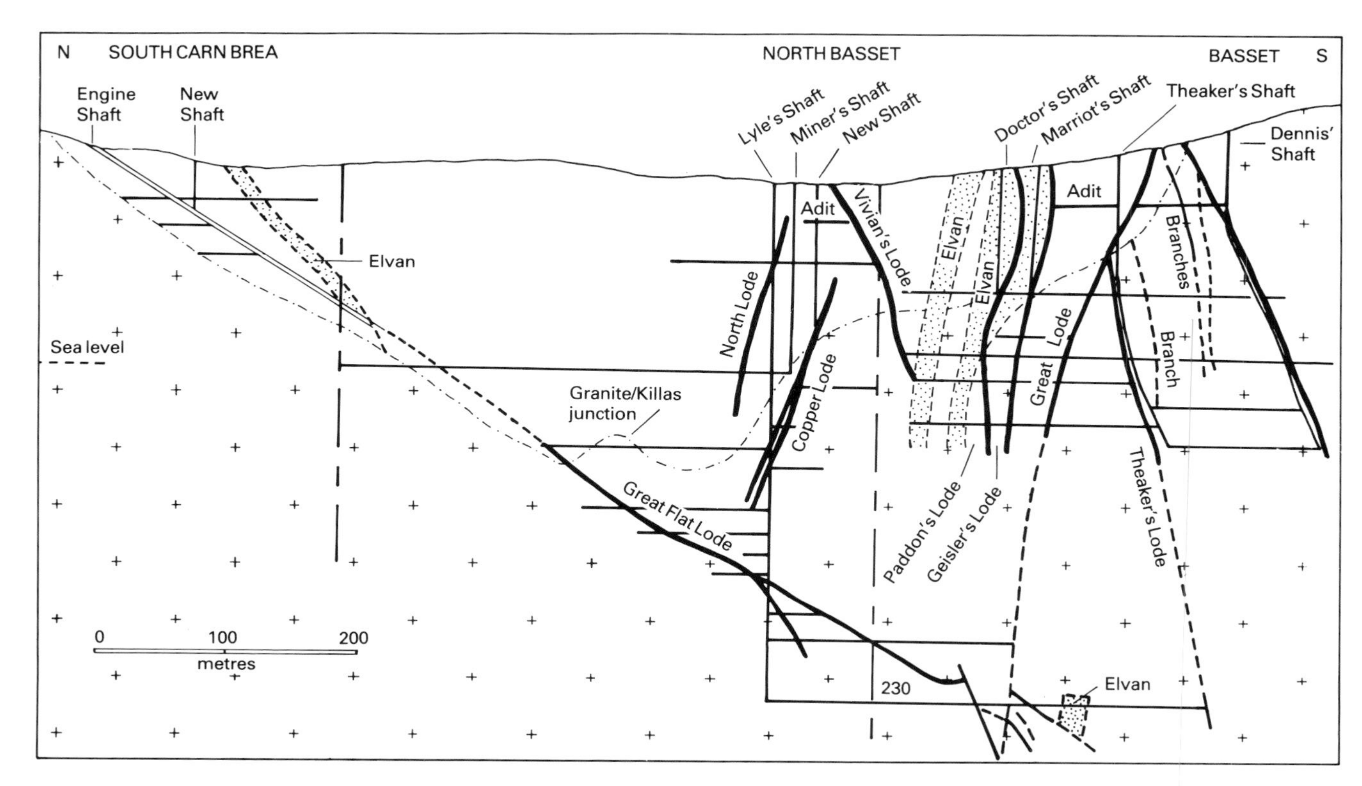

Fig. 17 Section showing the relationship of the Great Flat Lode to other more vertical lodes and the granite-killas junction within the Carn Brea-Basset mining area of Cornwall.

Fig. 18 Map of south-west England which illustrates the distribution of the main granite outcrops and their associated metamorphic aureoles, elvan dykes (thick lines) and principal lodes (thin lines) (after Hosking, 1964).

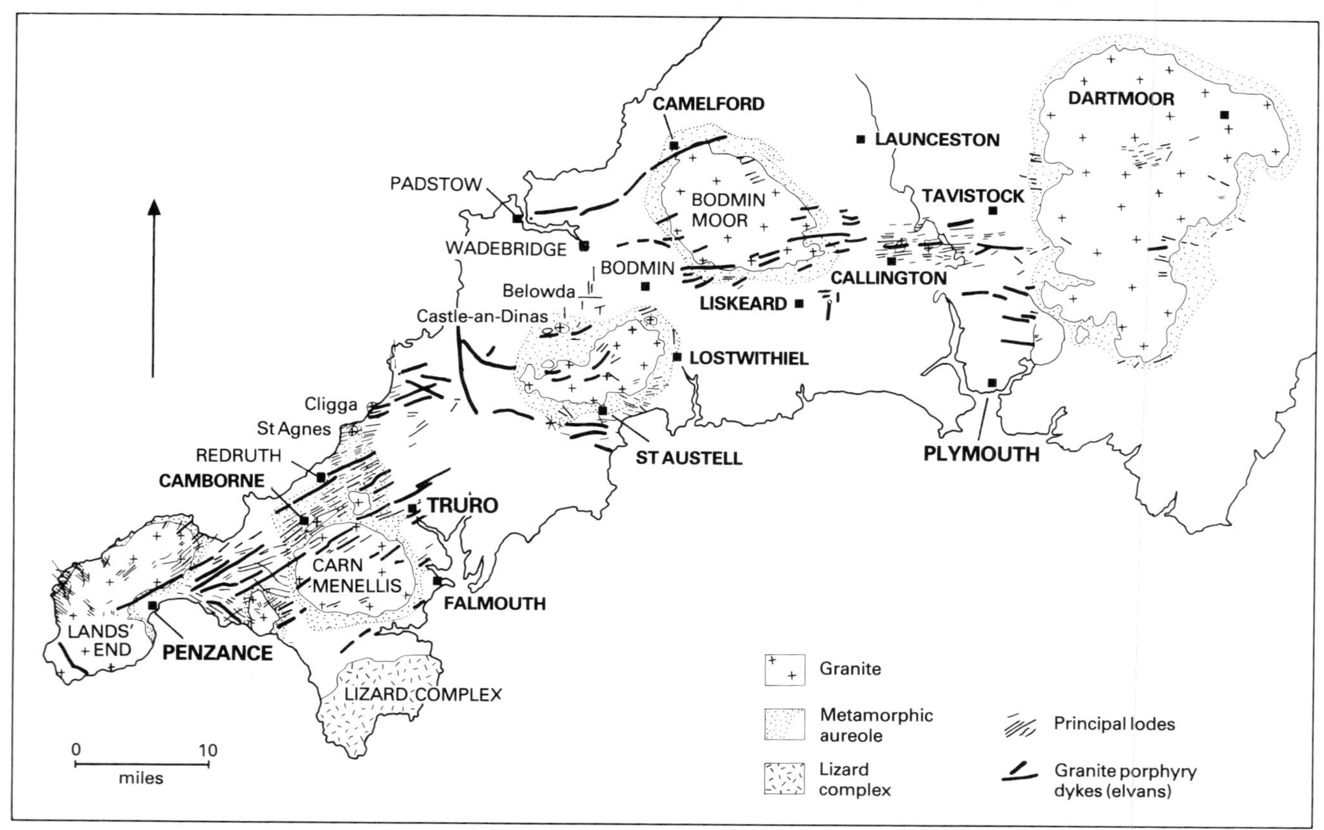

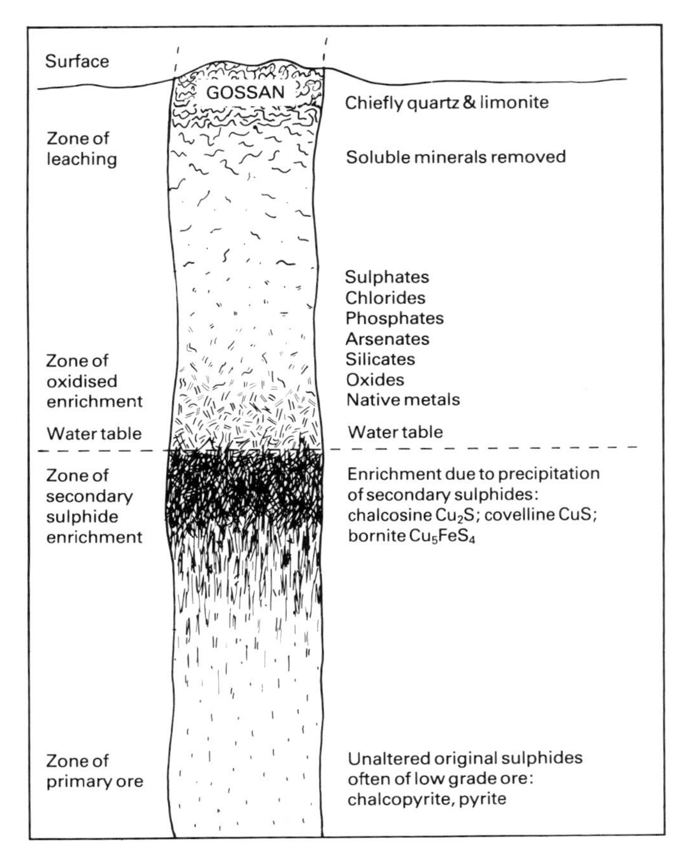

Fig. 19 Diagrammatic section illustrating enrichment of a copper ore body.

resistant components which may include cassiterite and even gold. Soluble sulphide minerals – such as chalcopyrite – were leached, transported, and redeposited above the water table as oxidized minerals. Below the water table, further copper precipitation as chalcosine and bornite (in the zone of secondary enrichment) occurred. The recorded depths of these zones of depletion and enrichment varied greatly, and the greatest extent of the oxide zone was at the Phoenix United mine where it reached 200 fathoms (1200 ft) from the surface (Hosking, 1950). We have no means of estimating the original vertical extent of these altered parts of the veins, the tops of which have been removed by erosion.

The influence of low-temperature alteration processes in the south-west may be gauged by the large variety and number of fine mineral specimens obtained from such secondary zones.

Zoning

It was long ago observed that tin and copper mineralization in south-west England tends to occur closer to the granite bosses than does, say, that of lead and antimony. Captain Charles Thomas, of the famous Dolcoath mine, successfully staked his reputation (and the shareholders' money) in the 1840s on his belief that rich tin ore lay beneath the copper. These and other observations were generalized (Dewey, 1925) into a now-classical interpretation, in which hydrothermal mineralization graded the metallic and gangue minerals into a series of roughly concentric zones or belts around the granite bosses. The theory relates the mineral zones, both laterally and in depth, to the falling thermal gradients between hot granitic magma and cool country rock. The major ore minerals are thus related to Lindgren's sequence for hydrothermal ore deposits (see fig.11). In general terms, this means that minerals crystallizing at higher temperatures (hypothermal, 300-500°C) and under favourable pressure conditions, such as cassiterite, wolframite, and tourmaline would occur nearest to the granite-country contact. The meta-sedimentary rocks would contain sulphides such as pyrite, chalcopyrite, and arsenopyrite; and in the successively cooler outer zones would be lead, zinc, and iron oxide minerals. Dewey instanced the Dolcoath mine, where zinc and copper were worked in the upper levels, copper down to 170 fathoms, copper and tin together from 170 to 190 fathoms, and tin below these to the bottom of the mine (550 fathoms) (fig.20).

The simple theory needed modification, to allow for changes in local conditions which caused overlapping of the zones, lack of certain zones in some areas, and even reversals of the order. For instance Wheal Jane, Kea, is a mine working a complex assemblage of cassiterite, sphalerite, chalcopyrite, and pyrite, and is situated beyond the metamorphic aureole of the Carnmenellis granite. And the occurrence of veins of appreciably different content, close together in the same mine, is by no means uncommon.

Fig. 20 Dolcoath mine, section of the main lode showing the relationship of the ore zones to the granite, the numerous shafts and the areas of copper and tin working.

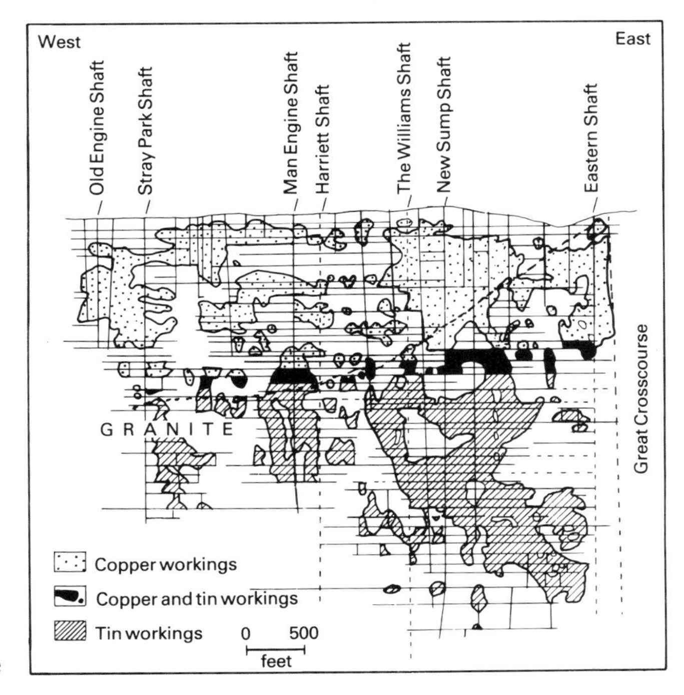

Dines (1934, 1956) recognised that the distribution of ore zones was not concentric about the main granite outcrops themselves, but about a number of 'emanative centres', which might be within the granite or the country rocks. He conceived the hydrothermal mineralizing fluids, which he believed to be of magmatic origin, as emanating from the hot core of the granite and passing along the principal joints that were established in the outer parts of the granite and in the country rock. Unfortunately, it proved difficult to identify these 'emanative centres' with any certainty; and the re-opening of the old New Consols mine, east Cornwall, in the late 1940s, was based on a wrong guess and so failed.

Recent studies of Cornish mineralization have shown that the popular form of the zonal interpretation is an oversimplification:

> 'all too often, accounts of primary zoning are so naive and such gross simplifications of the truth that, were the tin prospector to lay much store by them, he would be badly misled.' (Hosking, 1974:24).

Although the relationships may hold good in some areas, and in certain single lode structures, there are many other areas where zoning of this type is not evident. Several studies have shown that the component parts of a supposed zoning arrangement are in fact due to different phases of mineralization. Later veins often demonstrate 'over-printing', and possible reactivation of earlier mineralization patterns. It is difficult to reconcile the 'steady state' concept of thermal gradients with an episodic mineralization process.

There is also historical objection to the simpler forms of the zonal theory, which requires 'tin below copper'. Many of the early tin mines were abandoned when the mineralization turned to copper at depth:

> 'Tin in Cornwall seldom runs deeper than fifty fathom below the surface. Good copper is rarely found at a *less* depth than that.' (Warner, 1809:283).

It is not certain that all of this 'upper' tin was in the gossans. Nor is it explained why mineralization in the outermost zones (lead and antimony) should almost invariably occupy fissures with a quite different trend from that of the inner zone fissures. Many more age determinations need to be made before we can begin to be confident about the duration and sequence of the different phases of the mineralization process. A century and a half ago, Robert Were Fox (Fox, 1830; Henwood, 1843:445) made many observations on 'electricity' in the Cornish (and other) veins; it is surprising that there has been very little recent investigation along similar lines, for electrolytic processes might well contribute to the extreme localization and some textural features of metallic ore deposits.

Mechanism and summary

There seems little doubt that the primary tin–tungsten–copper mineralization came from hot, predominantly aqueous fluids, which distributed and redistributed a wide range of elements in several episodes, probably after the bulk of the granite had solidified. Systems of marginal fractures developed in the cooling intrusions and associated country rocks, enhancing their overall permeability. In both aureole and country rocks, these enabled heated 'formation waters' to circulate by convection and perhaps to mix with magmatic fluids. As the intrusions cooled, the composition and convection patterns of the fluids must have changed, infilling reactivated or new sets of fractures.

The mineral deposits of Cornwall and Devon have long been important in the understanding and study of hydrothermal fluids and ore formation, and it would seem that the Cornubian batholith and associated mineralization remains a rich source of problems.

> '. . . the plain truth is that we know singularly little about many of the aspects of this subject.' (Hosking, 1974:5)

Fluid-inclusion and stable-isotope studies are being used to study the nature of the mineralizing fluids but the actual source of the major elements tin, tungsten, and copper, and the reactions leading to the release and deposition of these elements, is still speculative. Some of the laboratory reactions leading to the formation of cassiterite have recently been reviewed (Eugster, 1986). Hydrothermal mineralization appears to have been active over a considerable time. If convection persisted as a major process, the energy to drive it must have originated in minerals containing radioactive elements, such as apatite, zircon, monazite, and uraninite.

The complex mineralization sequence, from widespread alteration and replacement to fracture-controlled processes, spanned the late magmatic and post-magmatic history of the granites; the whole being associated with various generations of tourmalinization, greisening, metalliferous mineralization, secondary alteration, and kaolinization.

Fig. 21 The Crowns section of the famous Botallack mine, Cornwall. The winding (upper), pumping (lower), engine houses are now mostly restored and preserved (see fig. 43).

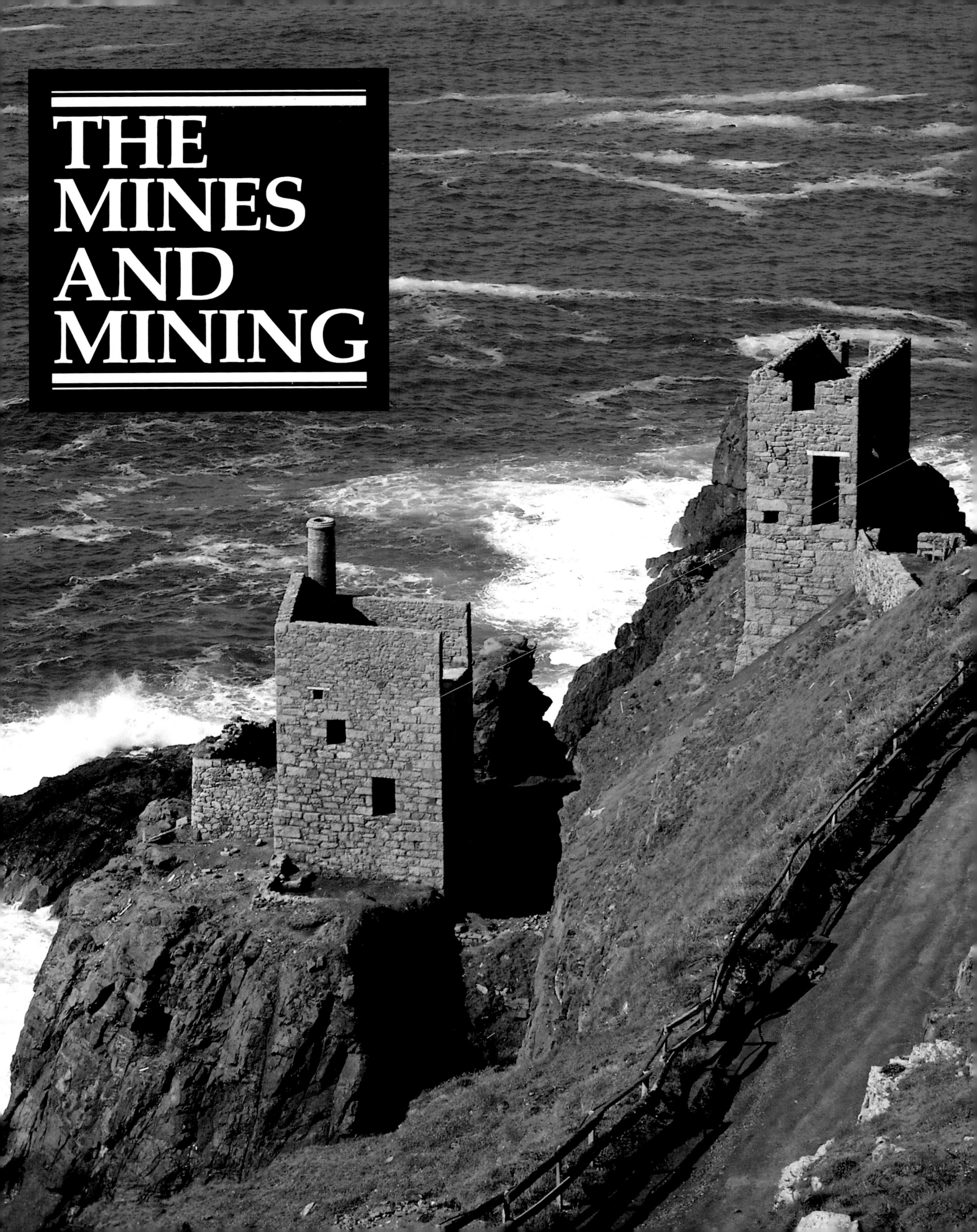
THE
MINES
AND
MINING

Mining history

Metalliferous mining in the south-west corner of England has been going on for so long, and has been so important in world trade, that even now, in its years of decline, the name 'Cornwall' instantly conjures up the words 'tin mines'. Probably no other area of the world has a comparably long history of mining activity, the traces of which may be found in almost every parish of Cornwall and in many parts of Devon. The mining fields of Cornwall, and to a lesser extent of Devon, are justly celebrated: tin and copper formed by far the greatest part of the area's metal output over the centuries; lead, silver, tungsten, and arsenic were also produced in some quantity, together with smaller amounts of zinc, antimony, nickel, cobalt, bismuth, uranium, iron, and manganese. Many individual ore deposits were extremely rich, though none was large by present-day standards. The Cornish miners had legendary skills and fortitude; they emigrated in times of depression at home, and their descendants may still be found in mining camps the world over.

The earliest workings for tin must have been among detrital material on hillsides, and in the alluvial gravels concentrated in the lower reaches of streams and rivers, from which early man recovered ore-rich pebbles and rock fragments. Tin-streaming has continued throughout history to a greater or less extent. Bronze Age artefacts have been discovered in Cornwall, and the first use of tin was probably coincident with the development of crude smelting methods for making tools and weapons out of bronze (an alloy of copper and tin). There were few sizeable deposits of tin in the ancient world, and the value of the metal for making bronze led to the establishment of trading routes between south-west England and centres of civilization by about 1500 BC.

The Greek historian Herodotus, writing in about 440 BC, gave the first account of the tin trade of western Europe; stating, however, 'nor have I any knowledge of Tin-islands [Cassiterides], whence our tin is brought.' On this, Hatcher (1973:10) comments that 'In a remarkable fashion these confessions of ignorance of the extremes of western Europe, and of the precise origins of Greek tin supplies, have inspired antiquaries to a wide search for the 'Tin-islands'', and gives references to accounts by later classical authors. Involvement of the Phoenicians in the Cornish tin trade of antiquity has been grossly exaggerated; no firm archaeological evidence has been found, and it is almost certainly mythical (Penhallurick, 1986: Ch.21).

The Bronze Age was succeeded by the new technology of the Iron Age (approx. 600 BC - AD 50), but even when iron became available it tended to supplement rather than replace bronze for some uses, and the established trading in tin continued. The Romans first settled in England in AD 43, and seem to have done a little prospecting for metals in Cornwall while their main needs for tin were supplied from their other provinces, such as Iberia. When these other sources approached exhaustion in the third century, and the development of pewter (an alloy of tin and lead) increased their need for tin, the Romans began to exploit the deposits of south-west

Fig. 22 Tin streamers at Red Moor, near Lostwithiel, 1900-1914.

England. During the Dark Ages, after the Romans had departed, tin working seems to have undergone a long period of recession. Its revitalization and organization came after the Norman conquest of AD 1066, and in 1201 King John granted the first charter to the stannaries, under which the tin workers were granted numerous privileges (Lewis, 1908; facsimile in de la Beche 1839, App.A).

Information on tin production after the conquest is contained in the 'Pipe Rolls' and other old taxation records. For a time in the Middle Ages, the alluvial deposits (and perhaps outcropping veins) of Dartmoor, in Devon, produced more tin than Cornwall and supplied an appreciable part of the world's needs. Also in this early period, there was true mining for lead and silver in Devon, by royal charter. The mines at Combe Martin, in north Devon, produced silver from lead ores in the reigns of Edward I and II (1272–1302–1327); the first miners being brought from Derbyshire for the purpose (Lysons 1822, 1:cclxxxv, 2:136). The mines at Bere Alston, in south-west Devon, were worked at the same time and later reopened in 1811 (ibid., 2:41).

As the working of ore-rich gravels proceeded, shallow exposed deposits were worked by open pits or trenches. We have noted that underground mining began much earlier for lead than for tin; for the latter metal it was probably insignificant before the sixteenth century, and then only where rich veins outcropped on cliffs or hillsides. Evidence of workings by the 'old men' may still be seen around the St Just area, and at Cligga Head; the former, perhaps, the 'old deep tin works' where swallows wintered, referred to by Carew in 1602 (1811 ed., p.85). The development of 'lode-works', as distinct from 'stream-works' (to use the old terms), seems to have started in west Cornwall and progressed eastwards. Norden, in his *Topographical and Historical description of Cornwall* (completed by 1605), refers to many mines in Penwith and Kirrier (the two westernmost 'hundreds', or political divisions); but to hardly any in the rest of the county, roughly east of a line joining Falmouth–Truro–St Agnes.

Native copper has occurred in Cornwall, notably near Mullion, where it was supposedly first found by chance, around 1720 (Hitchins & Drew, 1824, 2:503); individual masses were large, however, and it is hard to believe that none was found and exploited in antiquity. The earliest attempts to work its compound ores date from the sixteenth century. Richard Carew, writing in 1602, was dismissive of copper and metals other than tin:

> 'Touching metals: copper is found in sundry places, but with what gain to the searchers I have not been curious to enquire, nor they hasty to reveal, for at one mine (of which I took view) the ore was shipped to be refined in Wales, either to save cost in fuel, or to conceal the profit . . . But why seek we in corners for petty commodities, whenas the only mineral of Cornish tin openeth so large a field to the country's benefit?' (op. cit.:21,25; Halliday, 1953:87,88)

Norden, however, writing at the same time, was much more informative:

> 'Not farr from thence [Pendeen, Land's End], in the Innland, is great store of Copper and Copper mynes; as about *Moruath* [Morvah], *Sener* [Zennor], and *Lalante* [Uny Lelant].' and 'Sener, or *St Sennar*, [Zennor] a parish vpon the north sea, where are Copper Mynes verie riche.' (1728 ed.:40,41)

This is scarcely surprising since Norden, as a royal surveyor,

> 'seems to have had full information that the Cornish copper-mines were rich, and therefore in his letter to King James I . . . intimates the expediency of a better inspection ...' (Borlase, 1758:205; Pryce, 1778:276).

The annual output of tin began to grow rapidly in Tudor times and, from some 450 tons at the end of the fifteenth century, it averaged about 1500 tons for the next hundred years to about 1715 (Pryce, 1778:x; Hunt, 1884:817). During this period, primitive methods for breaking rock underground, such as fire-setting in shallow mines, gave way to larger-scale operations when gunpowder was introduced for blasting in the seventeenth century. The depth of mines was severely limited while water and ore had to be raised in buckets; streams to drive waterwheels were rarely conveniently situated for the mines, and real progress only became possible with the application of steam power in the eighteenth century.

> 'The history of Cornish copper is as a mushroom of last night compared with that of its tin. Lying deep below the surface of the earth, it would be concealed from the enquiries of human industry, till such time as natural philosophy had made considerable progress, and the mechanical arts had nearly reached their present state of perfection.' (Warner, 1809:282).

This comparatively late start was not occasioned by any lack of abundance, but much more probably because the extractive metallurgies of tin and copper from their common ores are so very different. 'Stream tin' is largely cassiterite, with few metallic contaminants, from which the readily fusible metal is obtained by simple reduction, originally using charcoal. The lucky accident, by which a mineral so unmetallic in appearance as cassiterite first came to be recognized as tin ore, is lost in prehistory; in

Fig. 23 Copper, from the Ghostcroft mine, Mullion, Cornwall (see fig.40): a large mass, weighing 635kg, with a little cuprite and malachite. This specimen (BM 1985 MI.11495) is 7ft 6in. long, up to 2ft wide, and varies from 4 to 12 in. in thickness. Easily the largest known piece of native copper from Great Britain, it was mined in about 1847 and presented to the Museum of Practical Geology by the Adventurers of the Trenance Mines after they had displayed it with even larger specimens at the Great Exhibition of 1851. Ghostcroft mine is said to have been named for the flickering will o' the wisp lights, or 'miners' candles', seen at night by the locals. Also known as Trenance mine, or Wheal Unity (one of several of that name in Cornwall), it was first worked about 1820 and abandoned about 1860, because the sporadic nature of the mineralization made it unprofitable. It is situated near the edge of the Lizard serpentine, from which native copper was already well known when the Revd. William Borlase wrote his Natural History in 1758.
(Note: Photographed in 1985, with John Fuller (left) and Peter Embrey (right)).

Cornish legend, however, it has been credited to St Piran, patron of tinners. The most abundant ores of copper, on the other hand, are sulphides; and these require considerably more fuel for a relatively complex smelting process, involving oxidation by roasting followed by reduction. These differences also applied to assaying: 'The processes for assaying Copper Ores . . . are little known out of the Cornish assay offices, and have too long been kept profoundly secret . . . The method of assaying Tin Ore is very simple ...' (preface, Pryce, 1778). Developments in smelting and refining techniques were needed before the numerous copper lodes of west Cornwall could be recognized as being valuable.

> 'Notwithstanding these hints [by Norden], I do not find anything material going on here in Cornwall, as to the improvement of the copper-mines, till, about sixty years since, some gentlemen of Bristol made it their business to inspect our mines more narrowly, and bought the copper raised for two pounds ten shillings per ton . . . the yellow ore [chalcopyrite] which now sells for a price between ten and twenty pounds per ton, was at this time called *poder*, (that is, dust) and thrown away as mundic [pyrite]. The gains were . . . so great, that they could not long be kept secret; ...' (Borlase, *loc. cit.*).

Celia Fiennes, writing in 1695, encountered copper mines on her way from St Austell to Redruth and referred to the smelting of the ore in Bristol (Chope, 1967:127).

This marked the start of large-scale mining for copper in Cornwall, which was to last for two centuries. For much of that time, its value greatly exceeded that of the tin production; and for many years it supplied a sizeable fraction of the world's needs (fig.24). Serious competition came in the last third of the century, when the Parys mine, in Anglesey, Wales, was discovered in 1768. Until it was worked out by the end of the century, this single rich orebody, on a hill and mined largely by opencast methods, caused a world slump in copper prices. It is curious, by way of contrast with the present century, that this slump occurred against a background of two major wars, of American Independence and the Napoleonic wars. Cornish copper mining, which had expanded enormously, carried tin mining down too in its decline.

Fairly early in the nineteenth century, the Cornish mining industry recovered rapidly. Copper production was 40 per cent of the world's supply in the 1830s, and reached a peak of 13 274 tons (metal) in 1856 (Hunt, 1884). Yet again, however, around 1866, came another serious decline in Cornish copper mining; prices dropped disastrously, due to new mining fields such as those of the Great Lakes area of North America.

Fig. 24 Diagram of percentage output of copper ores by producing countries from 1803-1913 (after Hatch, 1921).

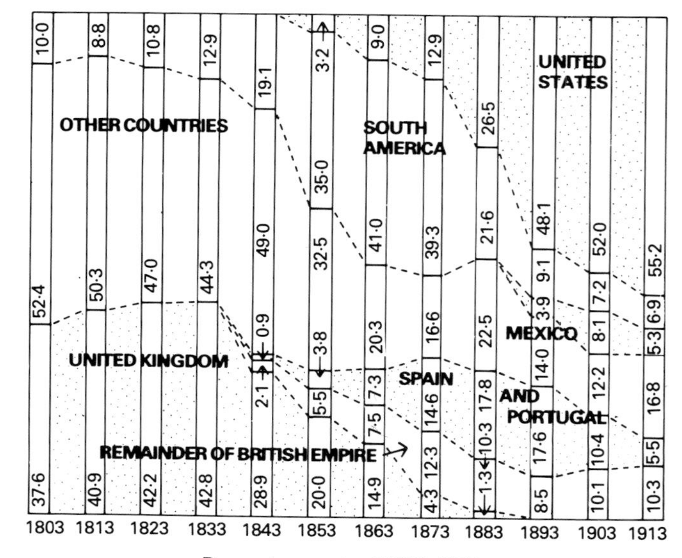

Devon once possessed a metal mining industry almost as diverse as that of Cornwall, but on a much smaller scale. In the 1880s, a hundred mines are said to have been at work, in addition to an unknown number of prospecting trials. In copper mining, Devon was an even later starter than Cornwall and produced only small quantities until, in the 1840s, there were considerable increases due to the rapid growth of the old Devon Friendship mine and the start of Devon Great Consols (Collins, 1912:262). In the years 1850–60, the latter became the richest copper mine in Europe, and later still was a major producer of arsenic from its arsenical ores. The first commercial arsenic production in Britain was in 1812, in the Carnon Valley area, with a considerable expansion in output around 1868 from Devon Great Consols. Standing chimneys and flues in the area are relics of this once-considerable industry.

Just as copper had earlier been found below tin, so it was also appreciated in several areas of the Cornish mining field that copper mineralization could be underlain at depth by tin reserves. With the passage of time, however, mine managements were no longer so specialized that they would automatically close a mine that ran from one metal to the other. As copper prices fell, the production of tin once more became important; the peak year for tin was 1871, when the region produced 10 900 tons, half of the world's supply. However, as with most predictions in mining, the actual richness of tin at depth proved to be variable from mine to mine: in some places the tin zone was found to overlap the copper zone, in others a barren zone existed between the two, and in yet others the tin zone was virtually non-existent. In financial terms, tin has actually played, overall, a lesser part than copper in Cornish mining. In their heyday the copper mines were more numerous, richer, and vastly more important to the economy of south-west England. Its longer history, aided in the past century by the disappearance of copper production, has made tin synonymous with Cornish mining in popular perception.

The slump in the mining industry of the 1890s led to the closure of all but a handful of mines, and once again miners went abroad to seek their fortunes in the new rich mining fields around the world. During the 1900–1914 period, with the advent of cheaper and more efficient machinery, the few remaining mines became more profitable but with a much reduced total output. The First World War created an increase in home demand for tin and tungsten, a temporary boom that was followed in the 1920s by a further recession; in 1921, only the small Giew mine, near St Ives, was producing tin. The famous Levant mine in the St Just mining field closed in 1930. The Second World War, again, produced a frenzied burst of metal production, and anti-tank armour-piercing shells, among other uses, increased the demand for tungsten. The tin–tungsten opencast at Hemerdon, Devon, was enlarged, and its mill processed ore trucked in from Carrock, Cumberland. The Castle-an-Dinas wolfram mine was small but rich; and the East Pool mine, Camborne, with government subsidy, was active until 1947.

After East Pool mine was abandoned, South Crofty and Geevor were the largest working mines left in Cornwall. New Consols mine, near the Devon border, was re-opened and the old workings were cleared at considerable cost; but expectations were not fulfilled, and it was soon abandoned again. Castle-an-Dinas mine ran out of ore at depth, and it too closed in the mid-1950s. Later in the 1950s, and in the 1960s, metal prices began to soar. Troubles in some tin-producing countries encouraged several companies to start prospecting in south-west England, and a strong revival seemed possible. There followed a few years in which Geevor mine expanded, Wheal Jane was re-opened, and new mines were started at Mount Wellington (near Wheal Jane) and Pendarves (near Camborne).

Tin prices continued at a high level into the 1980s, with heavy intervention purchasing by the International Tin Council. The success of manufacturing industry in finding cheaper substitutes, and over-production by countries with low costs, meant that supply exceeded demand and the stockpile grew. This could not continue, but instead of there being negotiated adjustments to the system it collapsed in spectacular fashion, in October, 1985. The ITC had overspent its funds, and could not honour its commitments, so the world price of tin dropped by half virtually overnight. ITC member governments failed to agree a rescue plan, after months of discussion.

The remaining Cornish mines, South Crofty, Wheal Jane, and the smaller Wheal Pendarves and Wheal Concord, cannot operate profitably with tin prices below about £6000 per tonne; Geevor is more expensive to run, and needs a price of about £8000. Nothing but a miracle can restore the world price to these levels in the short term. Government support has been agreed (August, 1986) for the first two, but Geevor has closed (October, 1986), so the Cornish tin mining industry is facing a bleak future. Matters would have come to a head sooner, perhaps, with a free world market or lower intervention price. The mining industry is notoriously vulnerable to swings in commodity prices, with its high overheads and the long time needed to bring or restore a mine to production. Alternating periods of boom and slump have characterised the history of mining in Cornwall and Devon, which in this respect only differs from other mining regions in having been much longer. The irony is that large tonnages of ore undoubtedly remain to be won, even after so many centuries of activity.

> 'It is, of course, a mere coincidence, although a curious one, that the last years of the sixteenth, seventeenth, eighteenth, and nineteenth centuries were all periods of great depression in Cornish mining.' (Jenkin, 1948:124).

It is a tragedy that this pattern has recurred in the twentieth century.

Methods and miners

Tin streaming – the early days

'Tin streaming', the whole process of working shallow alluvial deposits and concentrating the tin-rich pebbles and grains that are found in them, was an important occupation in the valleys and on the moors of Cornwall and Devon from the earliest times (fig.22). Over the centuries there grew up a complex mass of customs, relating to the informal rights of the 'tinners' to enter land while prospecting, to stake claims ('bounding'), to dig for tin and to divert streams, and so on. 'Bounds' were marked, on moorland, by cuts in the turf.

The tin produced was sufficiently valuable, both as a trading commodity and as a source of tax revenues, for these rights to be codified by statute, notably in the Stannary Charters of AD 1201 and 1305. As with most questions of law, there is no simple definition of 'stannaries'; roughly speaking, however, they were provinces of legal administration within the recognized tin-working areas. The legal definition of a 'tinner' was also uncertain, and changed over the centuries. There were four stannary divisions in Cornwall: Foweymore, Blackmore, Tywarnhayle, and Penwith. In each of these, a 'Stannary Court' took precedence over civil courts of law in most matters relating to the activities of the tinners, such as disputes over 'bounding', trespass and damage to property, tolls due to landlords, and contracts for the sale and purchase of tin ore and metal. The titular head of the system was the Warden of the Stannaries, a position held by Sir Walter Raleigh in Carew's time; most of the actual work, however, was delegated to the Vice-Warden and his officials.

The complicated division of rights, between the landowners and claim owners ('bounders'), led to many disputes when small adjoining mining 'setts', or leases, were being consolidated into larger mines in the eighteenth, nineteenth, and even in the present century.

Other civil rights accorded to the tinners included those of holding their own 'Stannary Parliaments', exemption from taxation (apart from that payable on tin), and exemption from military service. Under the charters, the stannaries of Devon became separated from those of Cornwall and paid lower taxes on tin than the latter. The Devon tinners could also prospect on enclosed (or private) land without prior permission, a freedom denied to the Cornish tinners.

There were designated 'coinage towns', to which all tin smelted in the district had to be taken (at fixed times during the year) for assay and for the payment of tax before sale; 'coinage' being the term for the removal of a 'coign', or assay sample, and the official stamping of the blocks of tin to indicate payment.

> 'The stamp, however, is said to afford no security for the goodness of tin sold abroad, since it is well known that in Holland every tin-founder is provided with it, and whatever his tin be, the inscription *block tin* makes it pass for English.' (Maton, 1797, 1:172).

The original coinage towns in Cornwall were Lostwithiel, Liskeard, Truro, and Helston; Bodmin was briefly one of them, in the reign of Edward I, and Penzance became one later. Tavistock was the most important of the three coinage towns in Devon. The last Stannary Parliament sat in Truro, in 1752; coinage was abolished in 1838; and, finally, the Stannary Courts were abolished in 1896, their jurisdiction being transferred to the county courts.

'Streaming', despite the many names given to the various stages of the process, was in its essence a straightforward undertaking, requiring the use of pick, shovel, and bowls. Before the advent of hard-rock mining, deer antlers and wooden implements sufficed. The work could be arduous, and life was lonely for the tinners and their families on the desolate, rough moorlands; but Carew refers to their 'lazy kind of life'. Ore from these remote areas was carried by horse-back to the nearest 'blowing-house' for smelting, and many examples of ancient ingots have been discovered (Penhallurick, 1986, Ch.26). In the blowing-houses at the end of the sixteenth century, the tin ore was laid on a charcoal fire

> 'blown by a great pair of bellows moved with a water-wheel, and so cast into pieces of a long and thick squareness, from three hundred to four hundred pound weight, at which time the owner's mark is set thereupon.' (Carew, 1811 ed.:41).

The blocks, made so much heavier than mere ingots in order to make theft difficult, were transported to the stannary coinage town for sale;

> 'A stranger will be very much struck, at his first entrance into Truro, to see the blocks of tin that lie in heaps about the streets' (Maton, 1797, 1:172; also, Hitchins & Drew, 1824, 2:645).

Gold was an interesting, but commercially insignificant by-product of streaming:

> 'The workmen take it as their perquisite, and sell the specimens to collectors. The value of the whole annual produce seldom reaches £100. The Carnan [now Carnon] and other *stream* tin works (for it has never appeared in any of the lodes) are the only places which produce it; where it is discovered in grains from the size of fine sand to masses (though very rarely) worth three or four guineas a piece.' (Warner, 1809:290).

The largest recorded nugget from the Carnon stream, at the head of Restronguet Creek on the Fal, was found in 1808; weighing about 56g, it is now in the Truro Museum.

Shallow mining

> 'The discovery of tin lodged in the bed of a river, leads to streaming; streaming leads to shoding, and shoding leads to the lode which is lodged in the bowels of the earth.' (Hitchins & Drew, 1824, 1:246).

'Shoding' was the process of tracing tin ore uphill in screes. As the richer alluvial deposits became depleted, search was made for the lodes in which they had originated, or for the more obvious outcrops of lodes in cliffs. Inland, the usual method of working shallow lodes was by 'costeaning', followed by sinking pits or exposing the lodes in 'coffins' (open trenches), maybe up to 50ft in depth. Rarely, instead of vertical lodes, 'floors' of mineralization were discovered at shallow depth; one of the best examples of this was the Grylls Bunny deposit, above the cliffs north of St Just, where a series of mineralized floors separated by country rock were worked at an early period.

Stream working, cliff mining, and shallow pit or trench working went on side by side for several centuries. Deeper working of the lodes, however, required extensive timbering of trenches. It became more efficient to sink shallow shafts to recover further ore, and this led eventually to deep mining. Such early workings were little more than trials, with shafts limited to a few fathoms, and were often worked, abandoned, and repeatedly worked again with changes of owner and name until the sett eventually merged into a larger group.

Deep mining and drainage

By the end of the sixteenth century, some of the mines had already reached the remarkable depth of forty or fifty fathoms. Mining proceeded by the familiar method of sinking laddered shafts, either down the dip of the lode or vertically, with horizontal drives at various levels to intercept the ore. Ore was then extracted, by upward or downward stoping between the levels. Patchy distribution of the ore as discrete segregations is shown on old mine sections by irregularities in the stoping pattern, and the effect of structural controls on the richness of mineralization can be seen from such drawings. Away from the main shaft, many smaller laddered shafts (winzes and raises) connected the levels and aided ventilation. For exploration, or to reach parallel known lodes or mineralized ground, cross cuts were driven; to save effort in removing unprofitable rock, these tunnels were often narrower and lower than the main ore levels. The details of underground working varied enormously from one mine to another:

> 'There is scarcely any department of life in which the exercise of judgment and discretion is so much required, as in the management of the mines, where local circumstances must be allowed to dictate to reason. A description of a mine, however accurate, can amount to little more than a description of that mine alone; ...' (Hitchins & Drew, 1824, 1:613).

Whenever practicable, adits were driven into cliffs and valley sides to reduce the labour and cost of raising ore and water from deeper workings. Sometimes, drainage adits were shared between two or more mines. The most ambitious of these was the County, or Great Adit, started in 1748 and completed fifty years later; eventually, with a total of 30 miles of linked tunnels, it drained forty-six mines in the Gwennap area, to depths of 40 to 80 fathoms (Henwood, 1843:89*; Collins, 1912:210; Jenkin, 1948:93, 1963 (VI)). Adits of this type may, on occasion, have disclosed valuable new lodes; but more frequently the cost of driving them must have put the profitability of a venture at risk.

> 'For conveying away the water they pray in aid of sundry devices, as addits, pumps, and wheels, driven by a stream, and interchangeably filling and emptying two buckets, with many such like, all which notwithstanding, the springs so encroach . . . [that] . . . they are driven to keep men, and somewhere horses also, at work both day and night, without ceasing, and in some all this will not serve the turn. For supplying such hard services they have always fresh men at hand.' (Carew, *ibid.*:38).

Some of these methods remained in use nearly two centuries later. From mines of modest depth, water was raised in stages by 'rag and chain' pumps; bundles of rag, attached to a loop of chain, moved in pipes of 3 to 5 inches bore and acted as a series of pistons to lift the water. Although usually operated by teams of men, in conditions of extreme hardship (Pryce, 1778:150), when the water supply permitted they were driven 'much more effectually and frugally by small water-wheels' (Borlase, 1758:171); as, indeed, they had been in German mines of the mid-sixteenth century, in a manner described and illustrated by Agricola. Deeper mines required lift or bucket pumps, also driven by water wheels; these continued in use long after the introduction of steam power, until they were replaced by pumps of the plunger pattern from about 1810 onwards.

Wherever it was available, water power was both cheap and convenient; but Cornwall's streams and rivers are, for the most part, poorly situated with respect to the mining areas. The small river Kennal provides an example of intensive utilization; in less than six miles, from its source in Wendron parish, it drove 39 water wheels for grist mills, hammer mills, paper works, gunpowder mills, wool-processing and spinning machines,

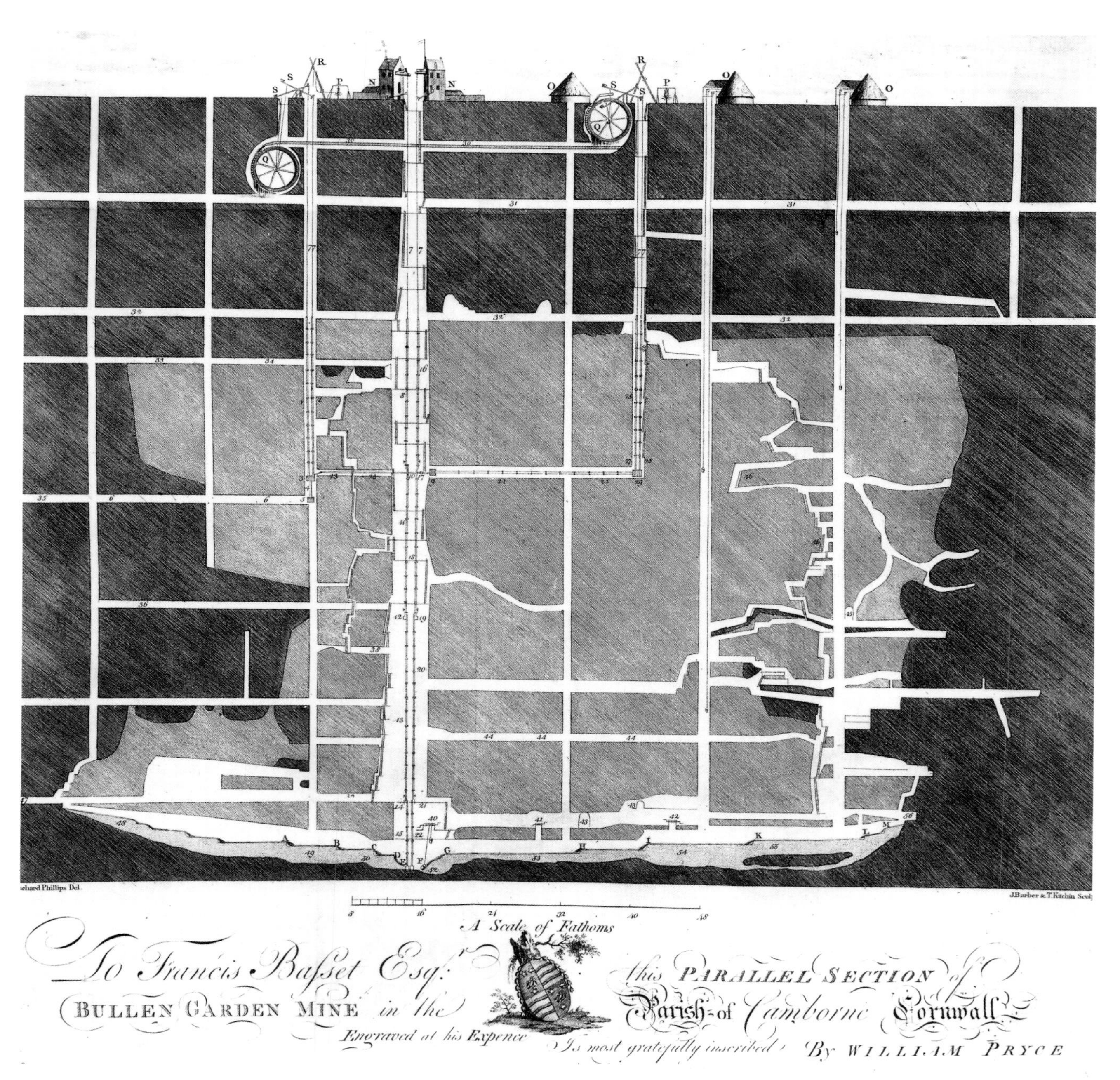

Fig. 25 Layout of the Bullen Garden mine, Camborne, Cornwall from Pryce (1778). NN = fire engines, OOO = whims, QQ = water engine wheels, SS = water engine bobs.

pumping and other operations at Wheal Magdalen, Ponsanooth, and, finally, boring mills, saw mills, lathes, and other engineering requirements at Perran Wharf (Hitchins & Drew, 1824, 2:621).

For driving stamps and other machinery at mines (and quarries), more rarely for pumping, water wheels continued in use well into the present century, long after the general introduction of steam engines and even electric motors. The earlier wheels were 12 or 15ft in diameter, later ones 40ft or more; Pryce (loc. cit.) refers to one of 48ft diameter at the Cook's Kitchen (copper) mine, where a few years later there was one of 54ft diameter working underground (Maton, 1797, 1:238). The largest wheel on record, at Boswedden mine, St Just, was 65ft in diameter (Barton, 1968:180). In order to drive these water wheels, and for other purposes, surface workers became skilled in diverting streams into artificial watercourses ('leats'): in an extreme example, water from the drainage adit of the Pednandrea mine was taken 6 miles by leat to Wheal Unity, Gwennap, for ore dressing (Jenkin, 1948:109; 1962(III):14; 1963(VI):11). Mine carpenters found constant employment, keeping the pump rods and laddering in repair, maintaining leats, and timbering the underground levels in areas of poor ground.

Fig. 26 Illustration of Cornish Stamps from Williams' Perran Foundry Co. catalogue.

Fig. 27 South Crofty mine, early this century, showing the old steam driven Cornish Stamps.

6

WILLIAMS' PERRAN FOUNDRY CO.

ELEVATION OF PUMPING ENGINE.

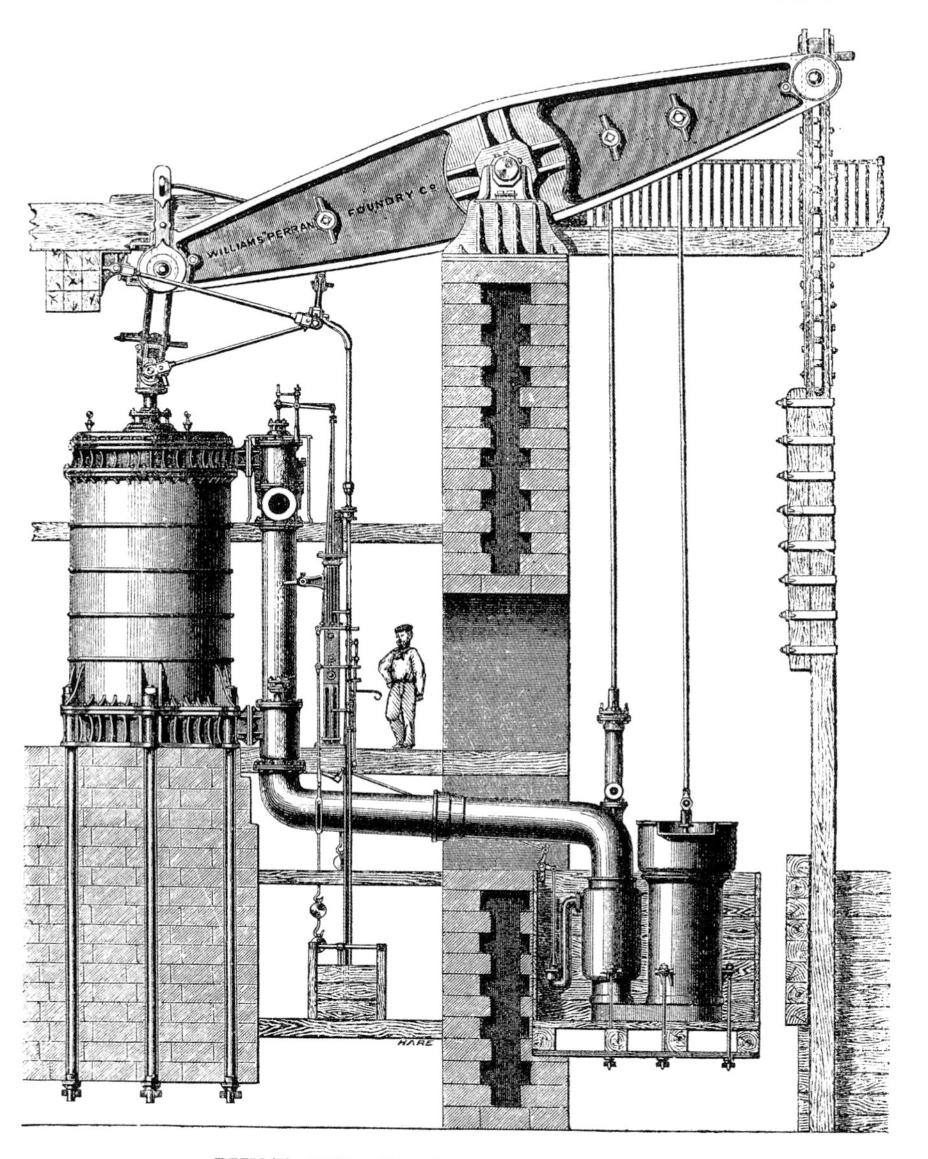

BUILT FOR ST. DAY UNITED MINES.

This Engine, made by us for the **St. Day United Mining Company** (now called the Poldice Mines), is a good example of the Cornish Engine as at present made, and gives a satisfactory idea of the general proportion of these **Engines.** It has been working under a load of 126,000 lbs. for the past eight years, and during the winter time has been kept continuously at work for six months, at an average rate of nearly eight strokes per minute, doing its work to the satisfaction of all concerned. **Estimates** on application.

August, 1870.

Williams' Perran Foundry Co.

PERRANARWORTHAL, CORNWALL,

AND

1 & 2 GREAT WINCHESTER STREET BUILDINGS, LONDON, E.C.

Fig. 28 A specimen page from Williams' Perran Foundry Co., catalogue illustrating Oppies 80 in. Engine at St Day United Mines, Gwennap. Built 1861-2 by the Perran Foundry.

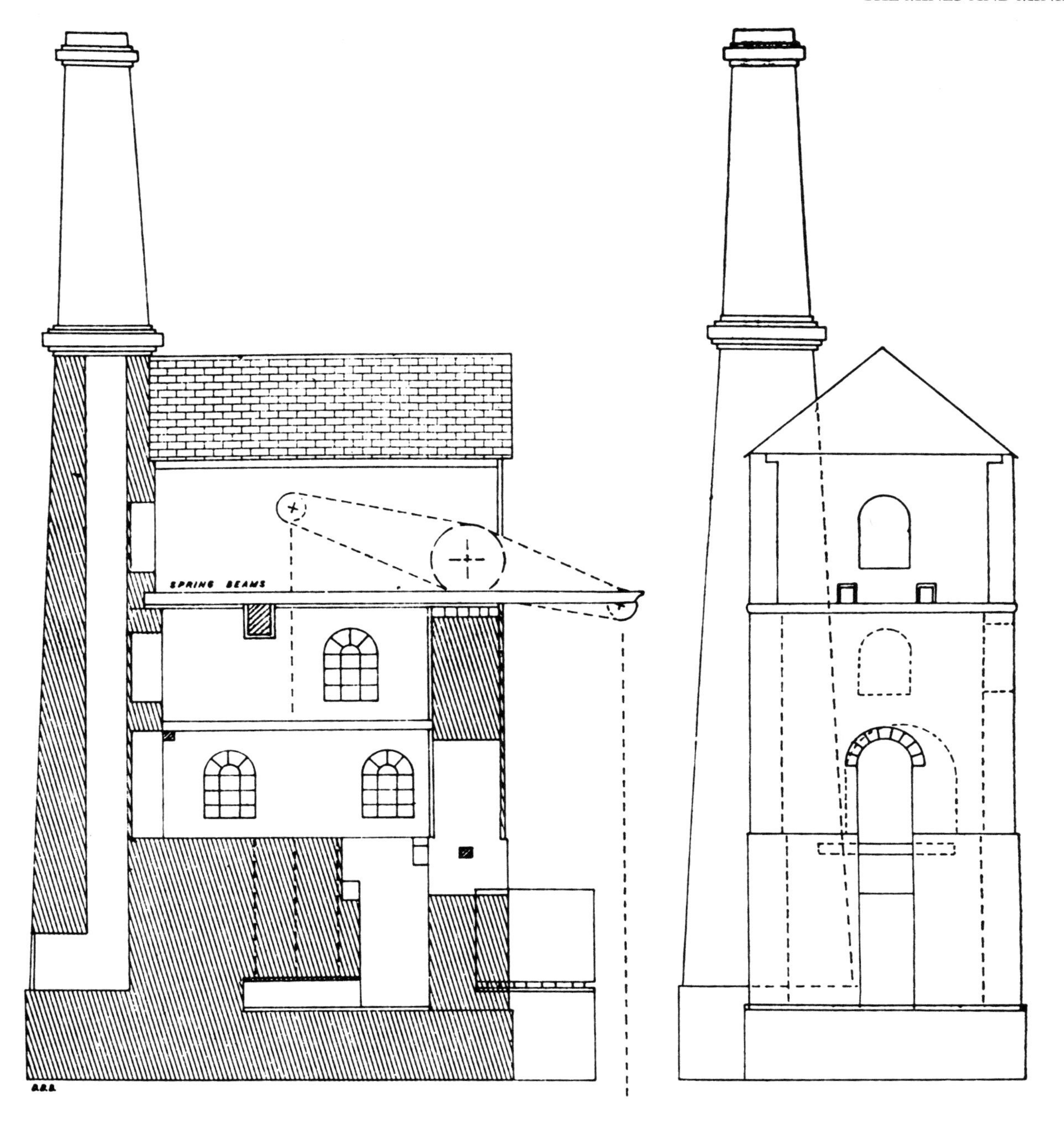

Fig. 29 Detail of engine house for Oppies 80 in. Engine 1861, this is a typical large three-storey Cornish engine house (see Barton, 1965). The heavy beam was pivoted on a strengthened (bob) wall and rocked up and down by steam power produced in a cylinder fixed to the engine house floor (see fig. 28). To the end of the projecting beam the attached rods worked pumps or powered winding machinery deep in the shaft (see fig. 30).

Steam pumping engines

Water power alone would have been wholly inadequate to permit the general deepening of mines in pursuit of copper, with the greatly increased requirements for pumping. The first steam engines for pumping water were invented and made by Thomas Savery in the first years of the eighteenth century, but although they enjoyed a limited degree of success in raising water for a few tens of feet, mostly for domestic purposes, they were of little use in dewatering mines; a Savery pump is supposed to have been tried at Wheal Vor, or perhaps another of the Godolphin mines, at this time. Savery's pump had no moving mechanical parts other than its valves, and the first practical engine using a piston in a cylinder was installed at a Staffordshire colliery by Thomas Newcomen in 1712. The considerable problems, for which Newcomen had to act as pioneer in finding solutions, included the machining of the cylinder bores, and linking the actuation of the valves to the movement of the piston.

Wheal Vor was also the site of the first Newcomen engine to be erected in Cornwall, in 1715, and by 1778 there were over sixty of them in the county. Newcomen's engines were, however, extravagant in their use of fuel for raising steam; and, while this was of little consequence when they were used at collieries in the Midlands and the north of England, the cost mattered greatly in Cornwall where there was little remaining timber, and coal had to be imported from South Wales. The performance of engines was measured by their 'duty', the weight of water raised through one foot by the burning of one bushel (94 pounds (Lean, 1839:140,146), or 84 pounds (Henwood, 1828:8)) of coal in the boiler; their size was measured by the internal diameter (in inches) of the steam cylinder. Many improvements were made by John Smeaton from 1765, increasing the 'duty' from about 5.5 to about 9.5 million pounds, and his most famous engine was the 72 inch installed at Wheal Busy, in Chacewater, in 1775. The single most important advance in efficiency, however, was James Watt's invention of the external condenser. A long and successful partnership with Matthew Boulton, a Birmingham industrialist, enabled Watt to exploit this invention; and the year 1778 marked the first introduction of their engines to Cornwall. While their patents lasted, Boulton and Watt did not sell their engines outright, but they collected a premium from the mine owners that was based on coal savings in comparison with a Newcomen engine doing the same work. They also made many 'rotative' engines, of smaller bore, first used widely in the textile mills of the north of England; when used in Cornwall as winding engines, the premium was based on their nominal horsepower since they did not raise water.

Rivalry between designers, and between those who operated the engines, in the quest for ever-greater fuel efficiency, made Cornwall the world leader in pumping engine development. Manufacture moved from Birmingham to foundries at Hayle, and at Perran Wharf on the Fal. Cornish beam engines were exported to all of the world's major mining fields, throughout most of the nineteenth century and well into the twentieth. Engines were often moved from site to site, as one mine closed and another started. The list of Cornish engineers involved in this work is long, including Bull, Lean, Trevithick, Hornblower, Woolf, Sims, and many others. The most efficient engine ever made was Hocking and Loam's 85 inch at Taylor's Shaft, on the Consolidated Mines in Gwennap, which in 1843 attained the exceptionally high duty of 105 million pounds. After the middle of the nineteenth century, however, the pursuit of high duties declined as it became clear how great were the strains and risks of breakage imposed by it on the engines. Surprisingly, only one descriptive account of the erection of an engine has survived (Francis 1845:41-57).

Cornish engines were also used extensively in waterworks and other pumping operations, and visitors to

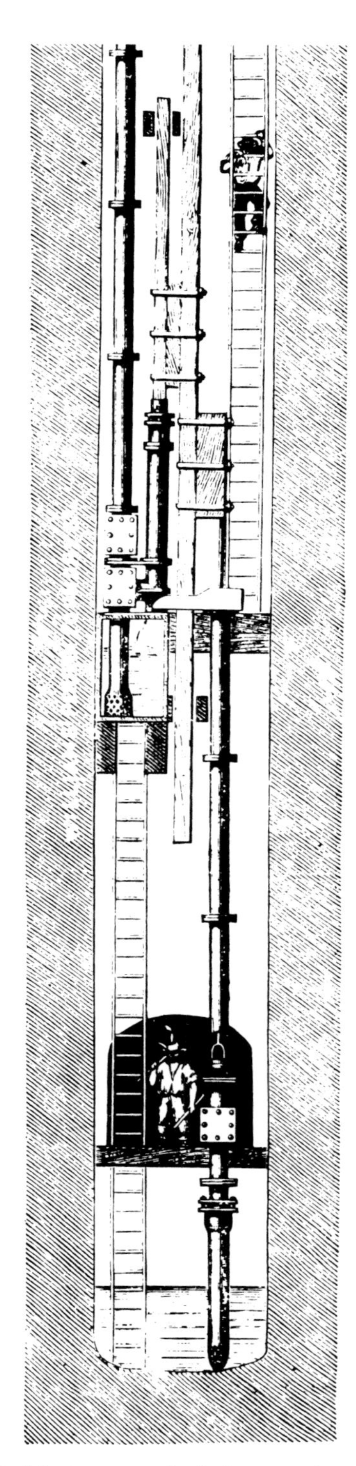

Fig. 30 Detail of drainage methods for typical pumping engine (Williams' Perran Foundry catalogue).

Fig. 31 In the 180 at East Pool Mine – the lode here has been worked away for a considerable length and height. The walls are fairly secure and do not require much support. The tramroad is maintained in good order, and ladder communication with the levels below and above is available. (Photograph by J.C. Burrow first published in *'Mongst Mines and Miners* in 1893. The description above is by his fellow author W. Thomas).

Fig. 32 Modern headgear of Robinson's Shaft at South Crofty mine alongside the preserved engine house.

Fig. 33 The importance of accuracy. Original pen sketch of Cornish mining characters drawn by George Seymour, over a hundred years ago. (George Seymour was a mining engineer and first president of the Institution of Mining and Metallurgy. Reproduced from the 'Man-Machine' by permission of the *Mining Journal*, London, 1977).

London can still (1986) see them at work, at weekends at the Live Steam Museum, Kew Bridge. The last large engine on a Cornish mine ceased work in 1955, and is now preserved at South Crofty. The largest engine ever built has also been preserved *in situ*, in Holland, after running from 1845 until 1933 in draining the Haarlem Meer; made by Harveys of Hayle, its cylinder had the enormous bore of 144 inches.

The remains of many of the stone buildings which housed these engines are familiar features of the Cornish landscape, although less common in Devon. The boiler houses stood alongside, and were less sturdily built, so that few remain and leave only their chimneys which were usually built onto the engine houses; above roof level, the chimney tops were characteristically of brick, for ease of handling beyond the scaffolding. The outer walls had to be sturdy, to carry the pivots for the heavy cast-iron beams, weighing up to 70 tons, which in the nineteenth century came to replace the earlier wooden beams. Working loads were all downwards, and in the later plunger pumps the engine acted only to lift the massive pump rods against gravity: the weight of the pitwork, on the 'outdoor' stroke, was enough to force the water to the drainage adit.

The miner at work

Underground mining conditions were extremely dismal, dark, damp, and dangerous. Perhaps the worst factor in the early days, even before the introduction of gunpowder and the poisonous products of its explosion, was the bad air from lack of ventilation; effective systems of forced ventilation were slow to develop, and working conditions were hot and oppressive. A graphic description of an underground tour of Poldice mine in 1791 conveyed the extreme unpleasantness very well (Clarke, 1793:90-98; Jenkin, 1948:102). Sudden flooding of the mine was a serious danger, particularly when there was a risk of breaking into poorly-charted abandoned workings nearby (Pryce, 1778:168). Such holing through into a 'house of water' cost many lives by drowning in the disaster at North Levant mine in 1867, and even greater loss of life at the nearby Wheal Owles in 1893.

> Using a 'pickaxe of iron about sixteen inches long, sharpened at the one end to peck, and flat-headed at the other to drive certain little iron wedges where-

> with they cleave the rocks', the going could be so slow that 'a good workman shall hardly be able to hew three feet in the space of so many weeks.' (Carew, 1811 ed.:35,37).

Light for working was provided by candles, attached to the walls or to the miner's hardened felt hat by a lump of sticky clay; spare candles were hung from his belt.

> 'In the *Loade* workes they worke all by lighte of Candle, and therfore harde to follow the Loade; which goeth somtimes sloping, sometimes directe, and oftentimes they loose it, and are muche trowbled to finde it againe.' (Norden, 1728 ed.:13).

The annual consumption of candles grew very considerably, amounting to 1 344 000 pounds in 1827 (Macfadyen, 1970:30). Acetylene lamps and, later, electric lamps, were only developed in the present century.

Gunpowder was first introduced to Cornwall, for underground blasting, in or about 1689. Accidents were common, greatly reduced only by the introduction of safety fuse in 1831; a plaque beside the main road at Tuckingmill, Camborne, still commemorates the inventor of the fuse, William Bickford. Even with the use of gunpowder, brute strength was required to drill the shot holes, and in the narrow levels of the St Just district the miners bored single-handed.

> 'In most places [the miners'] toil is so extreme as they cannot endure it above four hours in a day, but are succeeded by spells; the residue of their time they wear out at quoits, kayles [skittles], or like idle exercises. Their kalendar also alloweth them more holidays than are warranted by the church, our laws, or their own profit.' (Carew, 1811 ed.:35).

Tonkin's footnote to this passage, written about 1739, gives his view that the miners' toil was far less extreme:

> 'they do not work one half of their month for their owners [!] and employers . . . when once a fellow has taken to work to tin, he shall hardly be persuaded to do any thing else, though it were to keep his family from starving.'

Whatever the truth of the matter, in respect of the actual time spent down the mine, the miner's work was unquestionably dangerous, and his health was commonly broken by his mid-thirties – if he lived so long. Surface workers put in a longer day, and the night-shift driver of a steam pumping engine around 1800 gave this account of his time:

> 'I drove the engine for ten hours, worked on the farm seven hours, and wasted the rest.' (Lawrence, 1898:341).

Steam power for pumping water and raising ore enabled the mines to be sunk deeper, and the development of reliable air compressors both improved ventilation and resulted in pneumatic drills and boring machines being introduced, from the 1870s onwards. In the century before this, however, the effort of climbing 'to grass' by many hundreds of feet of ladders, after a hard shift at the face, became more crippling by the year. In the early days, miners were sometimes 'let down and taken up in a stirrup by two men who wind the rope' (Carew, 1811 ed.:36), a method similar to that experienced by Maton at the Wherry mine (1797, 1:209). This method was only suitable for small numbers of men, even when small steam winding engines were introduced. The need to raise men more efficiently led, in the 1840s, to the adoption of the 'man-engine', a sort of mechanical ladder that was first used in the Harz region of Germany, in the previous decade. Long rods, sometimes in pairs, were fitted with platforms at regular intervals along their length; and, attached at the top to the beam of an engine, more rarely to a crank driven by a

Fig. 34 Going down in a kibble (see fig. 33 for reference).

water wheel, they moved up and down in the shaft. At the top and bottom of each stroke, the miners would step alternately between the platforms of adjacent rods, or between a moving platform and another fixed in the correct position to the side of the shaft. In this way, and according to the sequence in which they used the platforms, the miners could ascend or descend rapidly in several stages of 10 or 12ft per minute, depending on the stroke of the engine. A water-driven man engine at Fowey Consols, in 1851, made three strokes of 12ft each minute and operated to a depth of 280 fathoms (Earl, 1968:67). The savings in time and health were incalculable, and miners were enabled to enjoy a longer working life. Gigs operated by winding engines were introduced in the late 1860s, but some man engines continued into the present century; the last of them failed in 1919, at the Levant mine, killing 31 men.

Organization of labour

On the early organization of the mines, as for so much else, we are indebted to Carew:

> 'If the mine carry some importance, and require the travail of many hands, [it is given a name] and they their overseer, whom they term their captain.' (1811 ed.:34).

Norden's account was similar:

> 'Euery great worke hath an ouerseere, and him they call captayne; whoe appointeth euerie man his Worke, and prouideth frames of timber to subporte the concauities where neede requireth, and Pumps to exhauste the water out of the worke, and other engines for that purpose.' (1728:12).

As the working manager, the captain needed to be experienced in all aspects of mining engineering and drainage, surveying, and the ways of miners; and the success of the mine depended on his knowledge and that of his agents. In times of prosperity, the 'bal kappen' (bal = 'place of digging') would be an imposing figure, perhaps white-coated and tall-hatted. A large mine, with considerable plant above ground, would have separate captains for the surface and underground sides of the operation:

> 'The whole business of this vast concern [Dolcoath mine] is under the management of a purser, a principal captain, eight inferior captains, and an engineer.' (Hitchins & Drew, 1824, 2:141)

The individual team of miners on a particular task was known as a 'pair' (or 'pare'), regardless of their actual number, and the 8-hour shifts (sometimes 6-hour) they worked were termed 'cores'. The method by which a miner was paid depended on his experience and the work he was doing. Labourers would be on set wages, whereas 'tributers' would enter into periodic contracts to work for an agreed share (tribute) of the value of ore sent to the surface. The size of the share depended on the richness of the ore, and the difficulty of hewing it from the lode, less deductions for the cost of tools and services.

Another type of contract miner was the 'tutworker', who was generally employed on mine development, shaft sinking, and driving cross-cuts; the rate would depend on the type and nature of ground to be worked, at an agreed sum per fathom. The actual work done was measured by the captain and agreed (after checking) by the 'pare'; a joking, anecdotal report of such a check measurement ran:

> 'A shovel and a shovel-hilt, a swab and a swab-stick, two great stones, and my two feet, just azackly.' (Anon, 1898:486).

The tributer worked the mineralized ground and lodes, and his earning ability depended on the application of his practical knowledge in arguing the contract share. The high reputation of the Cornish miner rested largely on the tributer, and his individual skill and judgment in assessing the run of a lode, the nature of the mineralization, and the probable richness of the ore.

Raising and dressing the ore

When broken loose, the rock was roughly sorted by the miners and the ore transferred to the haulage shaft. Wheelbarrows sufficed at first, and in smaller mines, but increasing production led to the installation of rails and tramming. Barren rock (deads, or 'attle') was often backfilled into 'gunnises', empty spaces where working had ceased, and the remainder hauled to surface to be discarded on the 'burrows', or waste dumps. Ore was raised 'to grass' in 'kibbles' (buckets), powered by hand windlasses or by horse-driven 'whims'. The term 'whim' seems to be an abbreviation of 'whimsy', a fanciful idea, and has been attributed to John Coster, who was associated with the beginnings of Cornish copper mining and smelting (Hitchens & Drew, 1824, 1:612); but the device itself was much older, and was familiar in Agricola's day. Steam whims (or 'fire whims') were introduced in the later 1700s, and by the middle 1800s were sufficiently powerful for the kibbles to be replaced by much larger steel skips; accidents caused by the breakage of chain links were greatly reduced when steel cables were introduced at the end of the nineteenth century.

Once the ore was at surface, the dressing processes began. Stream tin had presented relatively few problems in its preparation for smelting, because much of the dressing was already done by nature: weathering and

Fig. 35 Man engine in Dolcoath mine, Camborne, Cornwall. The means of ascending and descending in the mine by stepping onto and off platforms of the reciprocating engine rod. Each miner has a burning candle in his hat for light. (Photograph J.C. Burrow, 1892.)

Fig. 36 Group of miners outside the 'dry' at Dolcoath mine, Camborne about 1900 (Photograph by J.C. Burrow).

stream action had largely freed its grains from quartz; had graded it to a fairly narrow size range, and concentrated it; and had removed most of the sulphides and arsenides, together with much of the wolframite. Stream tin usually produced very pure metal.

The first steps in the treatment of 'mixed' ore from the mine were crushing and sorting. The sorting process was largely typical of the copper mines, since segregated patches of yellow chalcopyrite or grey chalcosine were larger and far more obvious to the eye than the cassiterite grains in any but the richest of tin ores. Crystals of cassiterite were called 'corns', or 'sparks': 'The best Ore is that which is in sparks.' (Merrett, 1678:952). Most of the sorting was done by hand, the ore being broken into smaller pieces, inspected by the light of day, and worthless rock fragments rejected. The heavier work of 'spalling' was done by men, after which women and girls (known as 'bal maidens') did the lighter work of breaking ('cobbing') and sorting the smaller pieces. This made the copper mines even more labour intensive than the tin mines:

> 'The persons employed at Dolcooth [Dolcoath] mine, including men, women, and children, those who are above and those who are under the earth, amount to about 1600. . . . of all the Cornish copper works, Dolcooth is the largest; though many of them employ six hundred men, beside a large tribe of women and children.' (Warner, 1809:133,290).

Sometimes entire families were employed on the mine, and so were even more dependent on its prosperity than was the community at large. Labour conditions changed for the better after the Royal Commission of 1864 (Schmitz, 1983).

The various stages in the mechanized dressing of lode tin ore have remained, in most of their essential features, the speeded-up equivalents of the processes that produced tin placers in nature: the larger rocks are reduced in size, by progressive stages, to release the cassiterite grains from gangue; flowing water carries the lighter gangue from the denser cassiterite, at each stage of size reduction; and, when necessary, roasting replaces weathering as the agency for oxidizing the 'mundic' (the old, generic term for pyritic sulphides and arsenides). Only the separation of cassiterite from the equally-dense wolframite differs: wolframite seems seldom to occur in

Fig. 37 Bal-maidens used bucking hammers, which had flat-heads about $4\frac{1}{2}$ in. square, to reduce ore to a coarse powder. The 'mill' refers to the flat iron plate, about 2 in. thick and 1 ft square, on which the ore was crushed. Bucking was a process undertaken after 'cobbing' in which the bal-maidens broke up the ore into small pieces and removed waste using small hammers. (Pencil sketch by James Henderson, a Scotsman who came to Truro in 1853 and published a series of drawings in his paper 'On the Dressing of Tin and Copper Ores in Cornwall', *Proc. Inst. Civil Eng.*, vol. xvii, pp. 195-220 (1857-1858)).

tin placers, having perhaps been removed by its greater ease of weathering; artificially, the first effective method was Oxland's chemical process, introduced in 1844 at the Drakewalls mine, and magnetic methods have been used latterly.

Stamping mills for pulverising ore were introduced and improved before the end of the sixteenth century:

> '. . . once brought above ground in the stone, it is first broken in pieces with hammers, and then carried, either in wains or on horses backs, to a stamping mill, where three, and in some places six great logs of timber, bound at the ends with iron, and lifted up and down by a wheel driven with the water, do break it smaller . . . From the stamping mill it passeth to the crazing mill, which between two grinding stones, turned also with a water wheel, bruiseth the same to a fine sand: howbeit, of late times they mostly use wet stampers, and so have no need of the crazing mill for their best stuff, but only for the crust of their tails.' (Carew, 1811 ed.:39).

The ground ore was spread on turf and washed by the water from the mill, being then turned over by shovel until most of the lighter gangue was removed from the heavier 'black tin'. There followed further concentration, by settling in wooden tubs, drying, and smelting. The more efficient 'Californian stamps', developed for crushing gold ores, began to replace the old Cornish gravity stamps at the end of the nineteenth century.

Allowing for later mechanical developments, such as jaw crushers, ball and rod mills, buddles and frames, jigs,

Fig. 38 Dressing Cornish tin ore in the mid-eighteenth century. Then, as now, the ore was crushed to free the different minerals from each other, and cassiterite ('black tin') separated in stages by exploiting its greater speed of settlement in water; the lighter impurities, above the 'black tin', were progressively washed away. In this simplified layout (Borlase, 1758, pl. xix, fig. 3) the stamps were driven by a waterwheel, and the crushed ore separated in the 'pits', 'buddles', 'trunks', and 'frames' for the finest material, or 'slimes'. The final separation took place in tubs or 'keeves' (N), before drying, assay, and smelting.

shaking tables, and belt vanners (to mention but a few of the devices), the operations of a modern tin-dressing mill would be admired but nevertheless understood by an expert of four centuries ago. Heavy-media separations, and the methods of froth flotation, used in the present century for removal of sulphides from comminuted tin ore, were innovations too recent for concentrating the Cornish copper ores. No single ore-dressing plant removed all the tin, and tailings from the larger mills were treated by one or more independent plants situated in series downstream.

Over the centuries, each major improvement in ore dressing techniques has led to profitable re-processing of the low-grade ores discarded on dumps by earlier workers. The rate of destruction of old mine dumps for commercial reasons has increased greatly in the second half of the twentieth century, and the process has been aggravated by misguided ideas on 'improvement' of the scenery. Many dumps, among them those at Wheal Gorland, Gwennap, which had long yielded fine specimens to the patient collector, have now disappeared.

Financial management of the mines

Financing a mining operation remained a simple matter as long as the scale was small, and management was in the hands of working miners. As early as the sixteenth century, however, members of the Godolphin family and other landowners were taking an active interest in the tin mines on their estates as well as drawing an income from royalties on the sales of ore. Large sums of money became involved in the eighteenth century, with the increased profits consequent on the advent of copper mining combined with the greater costs of sinking, draining, timbering, and of raising and dressing the ore from deeper mines. The partners or shareholders in a venture were known as 'adventurers', and up to the end of the Napoleonic wars they were, for the most part, members of the local landowning gentry or businessmen who had made their money directly on the mines or indirectly by supplying their needs.

> 'Many and various were the people to whom these copper fortunes went . . . [list of families] . . . [but] not a lot has been written on their commercial careers . . . Their charities, their scholastic abilities and their genealogies occur in print again and again but little or nothing of the ramifications of mining and merchanting and smelting and the like that had been the means of raising most of them from comparative obscurity' (Barton, 1968:72).

Second only to the captain, the purser was the most influential of those running a mine. Appointed by the adventurers, he needed to be a man of considerable experience and tact and was the equivalent of a company secretary, running his operations from the 'count-house' or accounts office. Accounting on the 'cost book' system, until the middle of the nineteenth century and beyond, was very much a hand to mouth business; heavy items of expenditure, such as a new engine, were met by 'calls' on the adventurers; and profits were divided in full between the adventurers, generally at quarterly intervals. Since no balance of working capital was retained, a mine could close down without warning if the calls were refused. The count houses were often large and substantial, and for the reception of the adventurers and other visitors were comfortably furnished compared with other mine buildings; two of them, at the Basset and the Botallack mines, have in recent years been converted for use as public restaurants. In the early 1800s,

> 'a mine account was generally held on the mine itself – a very good regulation, where the adventure is local; after which a dinner was provided for such of the proprietors as chose to attend, at the expense of the mine, also a fair and good regulation, as they who attend the business certainly deserve some kind of compliment . . . These expenses, however, are very different from the extravagancies to which such things were carried formerly. Mining count houses were more like regular hotels than mere places for transacting mercantile pursuits.' (Henwood, in Burt, 1972:29-30).

Later, as local adventurers came increasingly to be outnumbered by outside shareholders, and financing became more speculative and impersonal, the old celebrations of good dividends and the festivities accompanying major events passed into folklore; accounting methods underwent many changes, until all mines eventually became private limited companies.

Cornwall today, and to a less extent Devon, has abundant remains of its hectic industrial past; but these relics convey only a flimsy impression of how extensive that activity was. At the height of their fortunes, around 1862, there were 340 mines which directly employed 50 000 persons (Williams' *Mining Directory*, cited by Jenkin, 1948:171), and there were countless numbers of supporting businesses and their employees. Almost any site around the coast and on the rivers that could safely be reached by boat handled some part of the busy trade of exporting ore and importing supplies. Many of these were linked to the mines by small mineral railways, which in turn joined the main lines which connected the previously-isolated peninsula to other industrial centres. South Wales profited considerably, from the coal supplied to run the pumping engines and from the copper smelters set up near Swansea to handle the Cornish ores. The engineering works and foundries of Cornwall exported their machinery worldwide.

FURTHER READING. Many thousands of pages have been written, on nearly every conceivable aspect of Cornish mining, and the account above is but the barest of outlines. The best general account of the fortunes of the mines, and of the lives of the miners, remains Hamilton Jenkin's *The Cornish Miner*. This work has been reprinted, as also have the classical treatments of the subject contained in Carew's *Survey of Cornwall*, Borlase's *Natural History of Cornwall*, and Pryce's *Mineralogia Cornubiensis*, and so should be readily available in libraries. Burt (1972) has edited a collection of short essays, written by George Henwood (cousin of W. J. Henwood) for the *Mining Journal* between 1857 and 1859. One of us (PGE) first encountered Cornish mining in Ballantyne's *Deep Down*, a story for boys written shortly after the North Levant disaster of 1867. Technological aspects, up to the late nineteenth century, receive thorough treatment in Hunt's *British Mining*; Earl's *Cornish Mining* (1968) provides an excellent account of both historical and recent methods. *The Cornish Beam Engine* and other works by Barton are recommended reading. The best (and most recent) treatment of the archaeological aspects of the subject is Penhallurick's very readable *Tin in Antiquity*. These, together with many other related works, are in the bibliography.

The mines

Over the centuries, there must have been several thousands of mines in south-west England; even Collins' extensive but incomplete listing, with accounts varying in length from one line to a page or so, runs to over 200 pages (1912:399-613).

In this section, there are brief notes on only a very limited selection of some of the more famous mines (and interesting specimen localities) of south-west England, chosen to illustrate both historical and mineralogical features; for fine specimens have been collected from many of the mines.

Before the eighteenth century, mines were generally small, unnamed, and referred to locally as the 'Bal' ('a place of digging'); a miner on his way to work would be 'going to Bal'. 'Bal maidens' have been mentioned above, and the opencast Hemerdon Ball in Devon is a later form of Hemerdon Bal. 'Huel' (later 'Wheal', sometimes 'Whele') was the more specific Celtic word for a mine, also used in Brittany (as at the lead mines of Huelgoat); there was a plural form, but it seems not to have been used in mine names. As a matter of good usage, 'Wheal' ought not to be employed indiscriminately: thus, we may speak of Poldice, or the Poldice mine, but not Wheal Poldice, because this form was not used when the mine was working. Rarely, either form was used: thus, Janes mine, or Wheal Jane; but the use of both 'Wheal' (or Huel) and 'mine' (as in, say, 'Huel Jewel Mine' (Fox, 1834)) is quite wrong. 'Consols', a short form of 'Consolidated mines', and 'United', indicate unions of several smaller mines (or setts) under a single management. For an interesting general account of mine names in the west of England, see Barton (1968:93-113).

We have divided the region, rather crudely, into five areas of interest. The counties are now, for purposes of local government administration, divided into districts, and some of these bear the names of the ancient 'hundreds'; but few maps show these divisions clearly. It has long been traditional, in Cornwall at least, for mines to be specified by the parishes in which they are situated; unfortunately, however, their locations and boundaries are not easily found on any but the largest-scale maps, and there is no single or readily-accessible guide to the changes that they have undergone. It is often important, nevertheless, to specify the parish because, although several mines may have had the same name (e.g. Wheal Unity, or Wheal Prosper, with five or more of each), there would rarely have been more than one in a given parish; Wheal Rock, of which there were two in St Agnes, is a notable exception.

1) West Cornwall: St Just - Penzance - St Ives - Camborne (Fig.40)

The extensively mineralized cliff sections near St Just, in the extreme west of Cornwall, form one of the oldest, if not the oldest, of the mining districts in Cornwall. In the sixteenth century, or earlier, miners worked some of the tin-rich lodes exposed on the cliffs by tunnels driven inland just above sea-level. As mining progressed, shafts were sunk into these lodes to improve access and ventilation; and, as pumping facilities became available, the shafts were extended below the level of drainage adits. Eventually, several of the mines were extended beneath the sea.

Fig. 39 Composite outline map showing the five mining areas considered in the text.

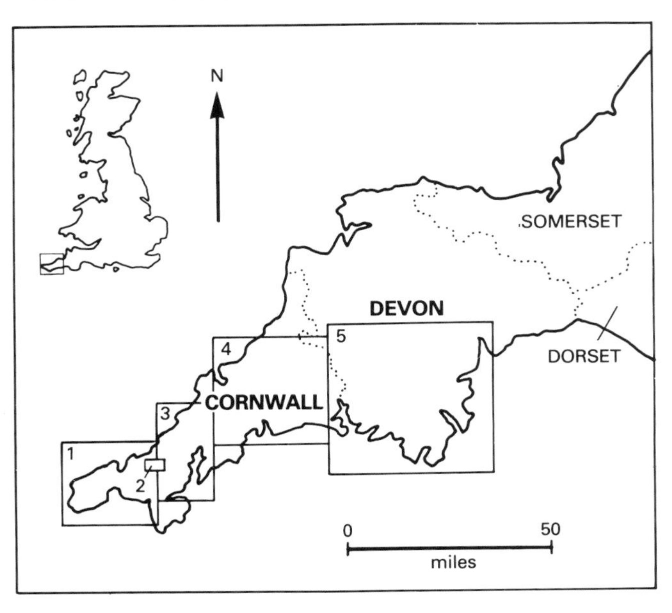

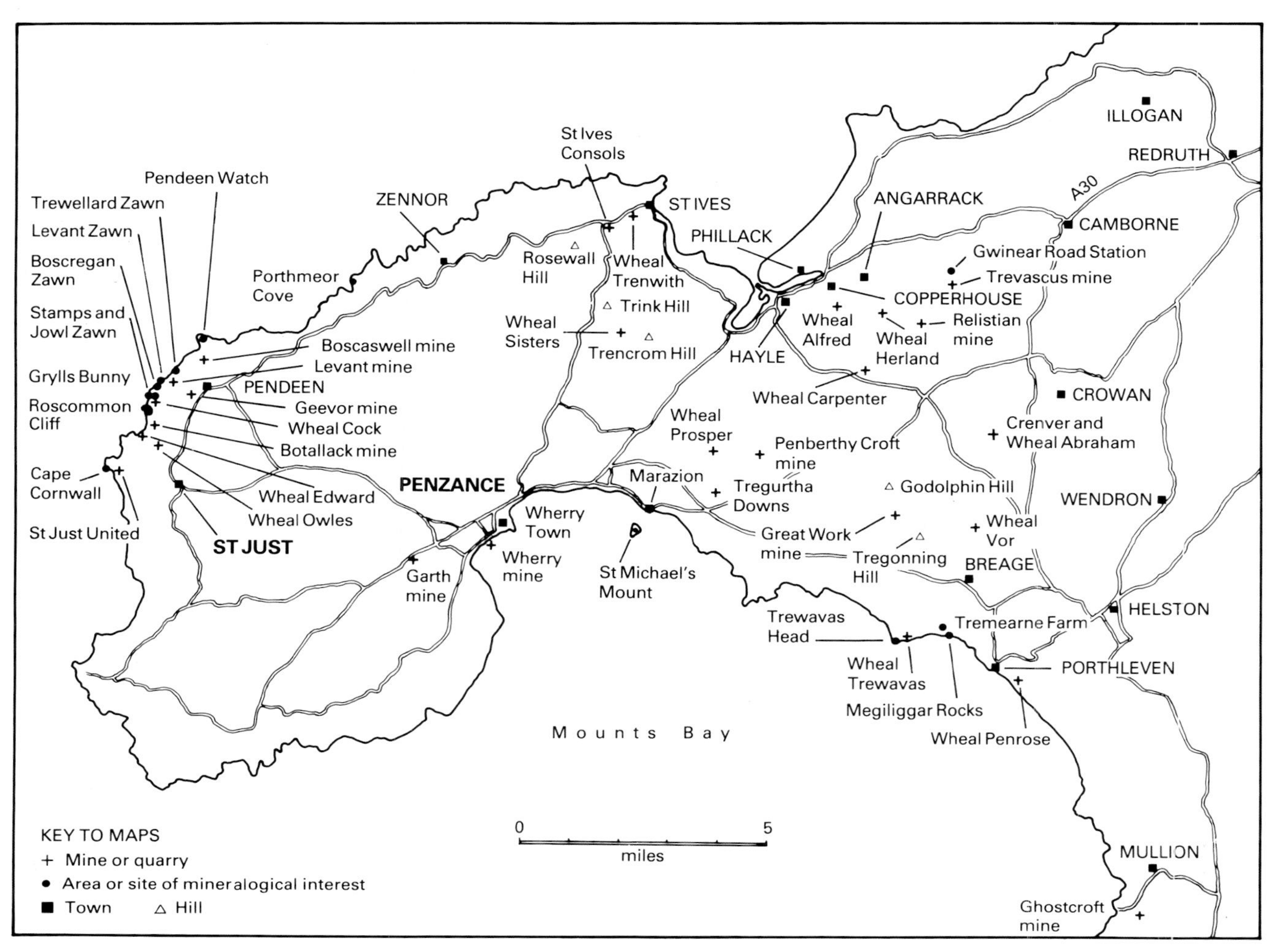

Fig. 40 Map of the St Just – Penzance – St Ives – Camborne area.

The lodes were worked within the Land's End granite, and were often found to be richest in the contact aureole, between the granite and the greenstone and killas country rock. The greatest concentration of mines is to be found in a strip about 4 miles long, running roughly from Cape Cornwall to Pendeen Watch; but there were many more, northwards and eastwards to Zennor and St Ives. It has been remarked (above) that Norden, writing in 1605, recorded copper mines in this area. The main copper ore, in both the St Just and St Ives mines, was chalcosine rather than chalcopyrite (Collins 1912:155). 'Zawns' are characteristic features of this coastline, steep-sided inlets where softer mineralized rock has been preferentially eroded by the sea.

Immediately south of Cape Cornwall is a group of old mines, known collectively as St Just United; small, and relatively shallow, they worked lodes (visible at beach level) towards and under the sea. A little inland, and north-west of St Just, is Wheal Owles, which at times included the extensive workings of Wheal Edward; in 1841, 200 people were employed and the mine had eight steam engines at work. In 1906, the setts were incorporated into those of Botallack mine, to the north. Two of the derelict Botallack engine houses are spectacularly perched on the edge of the rugged cliffs. Pitchblende, also zeunerite and related uranium secondaries, have been found at the nearby Wheal Edward.

Mining began here as early as 1721, and the rich tin (and copper) workings were soon extended beneath the sea. In 1866, the workings were 240 fathoms deep, had eleven engines, and some 500 employees. Its richness and dramatic setting have made Botallack one of the best known of Cornish mines. It was also famous for its 'diagonal' (Boscawen) shaft, down which the Prince of Wales (later King Edward VII) rode in a gig on a visit in 1865, a fatal accident two years earlier notwithstanding. Eventually, its workings extended some 2500 feet beyond the cliffs. Closed down in 1895, due to the low price of tin, it opened again in 1907; operations ceased in 1914 (Noall, 1972). Botallackite, found by Talling in 1865 and named for the mine, was almost certainly from the upper levels of the adjacent workings at Wheal Cock (Kingsbury, 1964:250).

North of Botallack, and close to the cliffs, was the equally famous Levant mine. Levant had a long and

highly profitable history as a producer of both copper and tin. The mine was known to be at work in the 1790s and, after restarting in 1820, became a steady producer of copper and of tin after 1852. It was still producing copper as late as 1910, and so was the last of the great Cornish copper mines. (Noall, 1970; on p.14 it is noted that 'Levant Zawn', on Ordnance Survey maps, ought to read 'Boscregan Zawn').

In the later years of working at Levant, tin was the more important metal. The mine reached a depth of 2000 ft (350 fathoms), and lodes were worked nearly a mile out under the bed of the Atlantic; two 'submarine' shafts connected deep levels. A man-engine was installed in 1857, to the 170 fathom level, and later extended to 266 fathoms; in the disaster of 1919, the rods broke loose and dropped to the bottom killing 31 men. After that, only the upper levels were worked until 1930 when the mine finally closed. The '40 backs' workings were very close to the sea bed; eventually, the sea broke through and flooded the upper levels.

Geevor mine is situated near the village of Pendeen, north-east of Levant. It had its origins in the nineteenth

Fig. 41 Cape Cornwall, Penwith, Cornwall overlooked by the stack of the Cape Cornwall mine viewed from the site of the St Just United mines. Remnants of the worked lodes can still be seen in the cliff sections at beach level.

Fig. 42 Levant mine, St Just. This famous mine was a major producer of both copper and tin. The exterior mining levels were driven some 350 fathoms deep and more than a mile out beneath the sea bed.

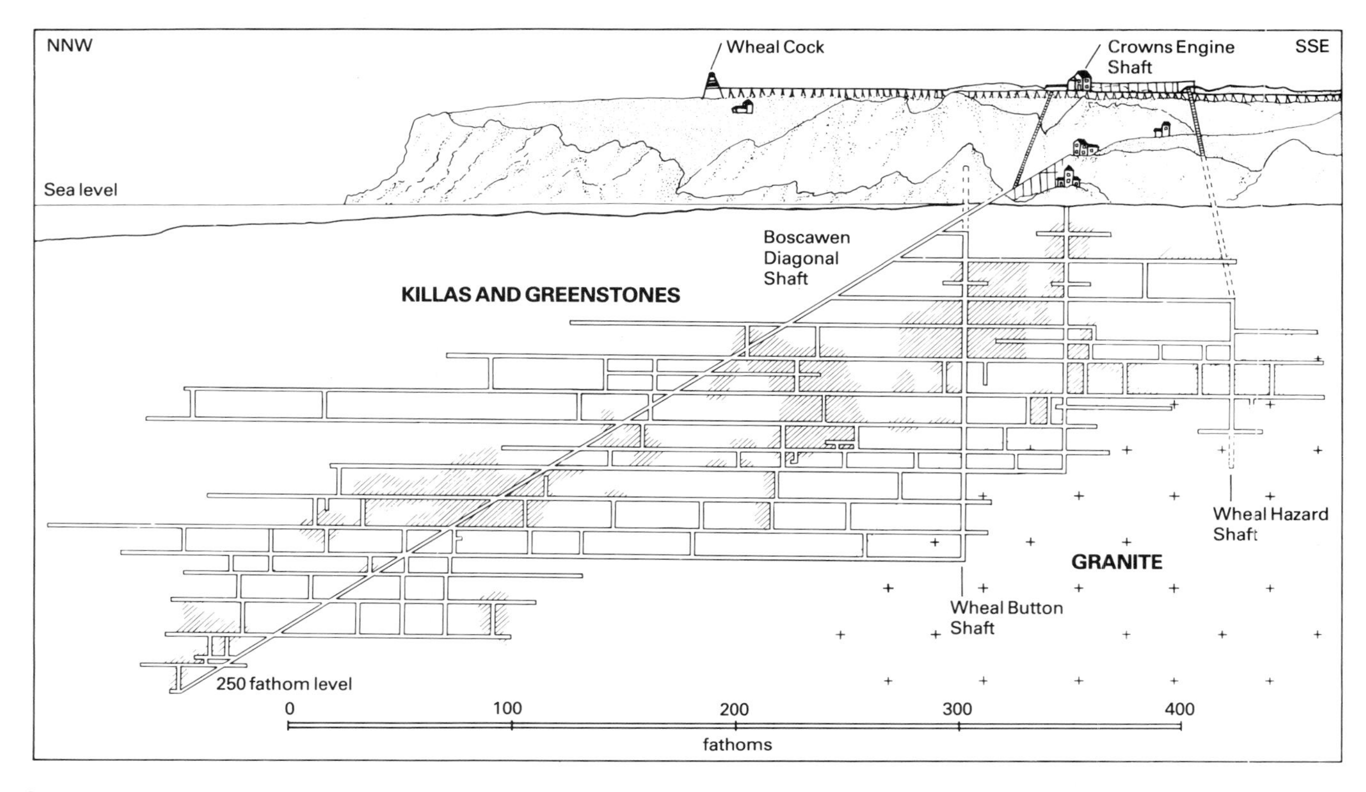

Fig. 43 Section through the workings on the Crowns Lode, Botallack mine, Penwith, Cornwall showing the worked areas and relationships of the ore bodies to the Lands End granite. Details of the preserved engine house standing above Crowns shaft are shown on page 15.

century with intermittent workings on a series of setts, known in 1837 as East Levant mine and in 1851 as North Levant. These survived the slump of the 1870s, and in 1906 a small syndicate revived them as North Levant and Geevor Ltd, the name changing to Geevor in 1911. A new shaft was sunk, and various old setts including the old Boscaswell and Levant mines were acquired. Many westward headings towards Levant had to be abandoned, due to the proximity of flooded old workings. By the late 1950s Geevor was in need of new ore reserves and, in order to recover those in the Levant section, attempts were made to seal the hole in the sea bed to the '40 backs'. A successful operation in 1969 led to the dewatering of the Levant lodes, and their development and exploitation. The collapse of the tin market in October 1985 has led to Geevor's near-closure a year later. There is a small mining museum, next to the disused Wethered shaft, but its future is uncertain; it should also be noted that the old Levant winding engine has been preserved, with restoration work done on it.

No account of the St Just area would be complete without mention of the mineralized veins in the altered rocks of the cliffs, many of them inaccessible except to an experienced climber. Perhaps of most interest to the collector are stokesite, from Stamps & Jowl Zawn (the exact locality for the type specimen is unknown, but was probably nearby); and axinite, sometimes as crystals to rival those from the classic French locality at Bourg d'Oisans, from Roscommon Cliff and near Trewellard Zawn (see p.84).

Although the remains of many isolated mines may be seen in this area, the next concentration of them, to the northeast, was centred upon St Ives. The richest mines were those of the St Ives Consols group, east of Rosewall Hill, which were worked mainly from 1818 to 1892, employing up to 400 people. Fine chalcosine specimens, different in habit from those of the St Just area, are well known to mineral collectors. Adjoining St Ives Consols, the dumps of the abandoned copper mine, Wheal Trenwith, produced commercial quantities of pitchblende in the early 1900s. Further to the south, on the slopes of Trink Hill, lay the Giew tin mine; well over a century old at the time of the 1921 slump, it had the sad distinction of being the only Cornish mine in production, and closed in 1923. East of Trink Hill, by Trencrom Hill, the Wheal Sisters group was worked until 1900; this was an amalgamation of several small copper and tin mines, including Old Tincroft and Wheal Mary.

Two miles west of Penzance, on the eastern side of the Land's End granite, was the Garth mine, which had been worked from the very earliest times; early in the nineteenth century it yielded some of the first 'wood tin' to be found *in situ*, and may have been the source of the masses 'nearly as large as a man's head' which were found built into roadside walls (Greg & Lettsom, 1858:358). Perhaps the most curious of all mines in Cornwall, however, was the Wherry mine. Situated some

Fig. 44 The Wherry mine, 1838, drawn and etched by Anne Margaretta Scobell of Poltair, Penzance.

240 yards out to sea, at Wherry Town on the southern edge of Penzance, the shaft was sunk on a mineralized dyke of 'elvan', or quartz-feldspar porphyry running parallel to the shore. Small amounts of tin ore had probably been dug here early in the eighteenth century, from a shoal exposed only at low tides, but the mine proper was testimony to the remarkable enterprise and perseverance of a poor, 57-year-old miner, Thomas Curtis. From about 1778 he devoted three years, working in the summer and when the sea permitted, to building a narrow wooden coffer 25 inches square and 20 ft tall to exclude the water, and to sinking a shaft; only then could he start regular production. Curtis died in 1791, when the mine had begun to make a good profit. Working was continued, and a trestle bridge was built between mine and shore. A steam engine on land operated the pumps by a system of flat rods carried on the trestle. The end came in 1798, when the coffer and trestles were demolished by an American ship adrift in a storm. The Wherry must be the only mine ever to have been destroyed by shipwreck; later efforts to re-open it failed. (Russell, 1949; Jenkin, 1962(IV)). In addition to cassiterite, some of which occurred as small but fine crystals, the Wherry mine produced copper and cobalt; 'smaltite' from this locality contains cobaltite and alloclase.

East of Penzance, to Porthleven and Helston and inland from Mount's Bay, there were several rich mines. Off Marazion, St Michael's Mount dominates the bay; collecting is forbidden, but mineralization is well displayed in the conserved outcrops on the island. Tregurtha Downs was the largest of a nearby group of tin mines, and very wet; it had a Boulton and Watt engine as early as 1778, and when it closed in 1903 its 80-inch engine (installed in 1854) was removed to South Crofty. The dumps around Daw's shaft, at the old Penberthy Croft mine, have yielded many specimens of bayldonite and a variety of good micromount material.

Also nearby, West Wheal Fortune and Wheal Prosper lived up to their names, as rich copper producers; the former was one of Cornwall's three greatest copper mines in 1725. The engine house of Wheal Prosper has been preserved; so, too, has that of another old copper mine, Wheal Trewavas, which worked lodes extending beneath the sea. Prominently sited on the cliffs of Trewavas Head, this is one of the spectacular monuments to Cornish mining. About a mile further east, on the shore below Tremearne farm, many pegmatite veins in killas are exposed on the Megiliggar Rocks.

The best of the tin mines in this area were around the granite bosses of Godolphin Hill and Tregonning Hill; the most northerly of these, Godolphin Bal, was for a very long time the largest of the operations: Norden recorded that it kept at least 300 people in continuous work, around 1600 (1728 ed.:45); and in 1765 it remained the most extensive tin mine in Cornwall (Jars, 1781, 3:194). It yielded good profits, and was still producing well in 1838, with two 80-inch pumping engines and 480 workers; but it closed before 1850.

Between the two hills was the important Great Work mine, with a long history of shallow working before deep mining started in 1825. To the southeast of Tregonning Hill, close to the granite and to the village of Breage, lay Wheal Vor. This mine had also been worked from early

Fig. 45 The cliff edge engine houses of Wheal Trewavas situated on the Tregonning granite at Trewavas Head, south Cornwall.

times by the Godolphin family, and is generally supposed to have seen the first use of gunpowder in Cornwall, in 1689; it is also reputed to have had a Savery steam pumping engine, in the early 1700s. With its own smelter, at its peak in the 1830s it was one of the largest of the Cornish tin mines, employing 1100 persons. It finally closed in the slump year of 1877.

On the north coast, at the narrowest part of the peninsula, is the port of Hayle, which grew considerably at the start of the nineteenth century to service the nearby Wheal Alfred (Hitchins & Drew, 1824, 2:553). It continued as an important centre of the Cornish engineering industry throughout most of the century, and the Harvey foundry exported pumping engines and other machinery for use around the world. The last of the Cornish copper smelters, at Copperhouse, Phillack, was connected to Hayle by a short canal; its operations were described in detail 'by a gentleman of high respectability' (*ibid.*), probably Joseph Carne, but it ceased work in 1819, shortly after the (first) closure of Wheal Alfred. The tin smelter at Angarrack, also nearby, operated from 1704 to 1880, and both it and Copperhouse depended on Welsh coal shipped to Hayle.

About a mile due east of Hayle station, on high ground, lie the extensive remains of Wheal Alfred (also known as Great Wheal Alfred). A very wet copper mine, it was extremely profitable from 1800 until 1815, but heavy expenses and mismanagement caused it to close in 1816 (Jenkin, 1959); it may also have been worked before 1800 (Collins, 1912:181,401). Subsequent attempts to rework it, from 1823 to 1826 and 1851 to 1862, resulted in loss. In the first of these attempts, however, many beautiful specimens of greenish-yellow mimetite were recovered, undoubtedly from a cross-course; and on a very few of these, agardite-(Nd) has been found as an associated species. Wheal Herland, about a mile further east and on the same main lode, was a much older mine and had a 70-inch Newcomen engine at work in 1746; a remarkably rich pocket at an intersection with a cross-course yielded 115 tons of silver ore, much of it native, in 1799-1800 (Hitchens, 1801). A mile further east lay Relistian, a shallow tin mine until 1719, which was reopened for copper at the end of the century; the lode seems to have been irregularly enriched in copper and tin ores (Jenkin, 1963(V):34). It produced specimens of an unusual and characteristic conglomerate, consisting of chloritic pebbles cemented by cassiterite, which may be found in many old collections.

About a mile to the south of Wheal Alfred and Herland, the dumps of Wheal Carpenter produced the first specimens of cornubite (Claringbull *et al.*, 1959). Trevascus mine, an ancient copper mine lying just to the south of the old Gwinear Road railway station, produced abundant specimens of fancifully-shaped chalcedony (Borlase 1758:122). Wheal Abraham, once the deepest mine in Cornwall at 240 fathoms (Carne, 1822:124), lay about a mile south-west of Crowan; in 1813 it produced some remarkable specimens of chalcosine, in crystals of nail-head habit (Sowerby, 1817, 5:Tab DXVIII) (see fig. 80).

2) Camborne - Illogan - Redruth (Fig.46)

The next large granite boss, to the east, is Carnmenellis (rarely, Carn Menelez); the smaller, elongated boss of Carn Brea lies to its north-west, south of the road between Camborne and Redruth. Mines to the east of the neighbouring boss, Carn Marth, are in the next section (3).

The area on and around the slopes of the Carn Brea granite was long regarded as the centre of Cornish mining, with 100 or more mines active at one time or other. The many large and famous mines worked a richly mineralized zone, about 3 miles wide by 4 miles long and running roughly WSW–ENE. Despite the area's former importance, and the widespread evidence of past mining activity, only the South Crofty mine and the smaller Wheal Pendarves remain at work. The Camborne School of Mines, no longer in its old buildings, is on the main road (A30) close to the old East Pool and Agar mine; and beside the Portreath road is the Tolgus Tin 'Streaming' Museum, which actively processes alluvial tin and dump material.

The Carn Brea area falls naturally into two major parts:

a) On the northern slopes of Carn Brea, a zone of famous mines straddles the main (A30) road. The familiar names Dolcoath, Cook's Kitchen, South Crofty, Tincroft, Carn Brea, and East Pool are among the mines in this group.

b) On the southern side of the Carn Brea granite ridge, towards Carnmenellis, the 'Great Flat Lode' was worked in South Condurrow, Wheal Grenville, and the Basset mines.

Many of these mines lay in the parish of Illogan, which was absorbed into neighbouring parishes by boundary reforms in 1934.

After the shallow tin was worked out, most of the mines of the area were originally worked for copper; later, and at depth, tin was also often discovered and exploited. Most of the finer tailings from the dressing floors of these mines was discharged into the Red River,

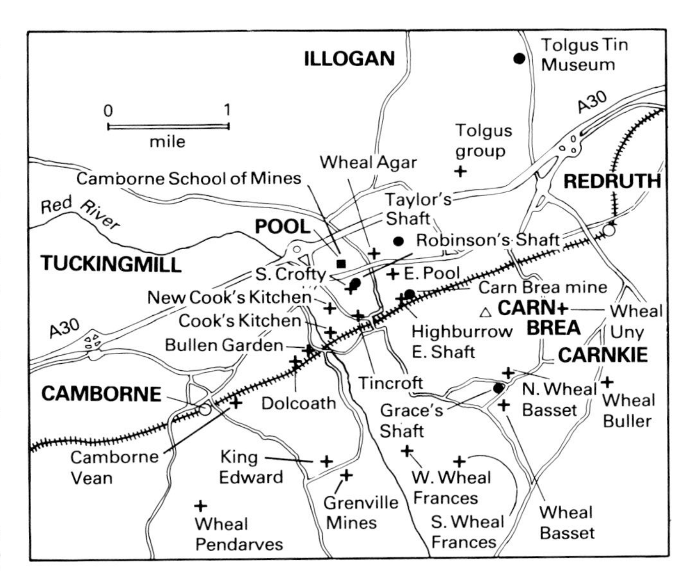

Fig. 46 Map of the Camborne – Illogan – Redruth area.

and was further treated by numerous small tin streaming works.

Dolcoath, with its bottom level eventually reaching 550 fathoms, was the deepest and perhaps the most famous of all the Cornish mines. As the first and outstandingly successful example of the daring application of a theory of ore zonation, by Captain Charles Thomas, it merits an extended account. The mine was situated on land belonging to the Basset family, at Entral, where Raspe set up an assay office in 1784. Worked for copper from the 1720s (or before), it incorporated many other properties including the Bullen Garden mine described by Pryce (see fig. 25).

The great copper slump caused the mine to close in 1787, but it was restarted in 1799 and soon returned large profits. The copper ores became depleted at depth, and in 1836 the lower levels were abandoned. The underground agent Charles Thomas was, however, convinced that rich tin ore lay below the copper zone; his appointment as captain in 1844 marked the renewal of deep mining, and the departure of the 'copper' shareholders who were not prepared to invest in his theories. Tin ore quickly became richer with increased depth, and the first tin profits were returned in 1853; Charles Thomas was succeeded as captain in 1867 by his son Josiah, who fully shared his faith in deep mining, and under their management Dolcoath enjoyed its second great period of prosperity, this time for tin. Other metals, such as arsenic, cobalt, bismuth, and tungsten were produced in smaller quantities, and in 1864 the mine had ten steam engines, seven water-wheels, a man engine, and employed 1200 persons. Captain Josiah Thomas introduced pneumatic drilling in 1876, and at surface replaced the old Cornish stamps by the more efficient Californian stamps in 1892. The lode was worked to a width of 30 ft in some places, presenting problems in tim-

Fig. 47 View of mining region on north side of Carn Brea looking east. Formerly in possession of the mining historian, the late A.K. Hamilton Jenkin. An etching, unsigned, undated, but probably *c.* 1930.

Fig. 48 The Camborne – Illogan mining area seen from the foot of Carn Brea, looking west to the hills of West Penwith in the distance (Photograph H.G. Ordish, June, 1934).

Fig. 49 Tuckingmill Valley, Camborne, Cornwall. Cooks Kitchen mine dressing floors in foreground, Dolcoath dressing floors behind observer; North Roskear engine house is on the skyline to the left with arsenic stacks cutting the skyline to the right (Photograph by J.C. Burrow about 1900).

bering. A 'run', or collapse, in 1894 killed eight miners, and caused a temporary closure; underground working finally stopped in 1920, with the total production of 80 000 tons of tin ore and 350 000 tons of copper ore. A fire at the count house, in 1895, destroyed many of the old records.

The very productive and important Dolcoath Main Lode was worked from Camborne Vean in the west, through Cook's Kitchen and Tincroft, to Carn Brea mines in the east, for an overall distance of nearly 3 miles and closely parallel to the main railway line. The Cook's Kitchen and the Tincroft mines were both very old, having started working for copper early in the eighteenth century; like many of the mines in this area, they probably produced shallow tin before this time. Both were very productive of both copper and tin ores, and highly profitable in their early years. Tincroft was incorporated into the Carn Brea mines in 1896. Old and extensive, these mines had been restarted for copper in 1831 and changed to tin in 1865; they finally ceased work in 1913. Fine crystallized specimens of cassiterite and copper sulphides were characteristic of this area; some of the Carn Brea fluorites were connoisseurs' delights, and Tincroft produced remarkable groups of pale-brown siderite crystals, also equant torbernite, and hematite.

Pool is on the main (A30) road between Camborne and Redruth, and lent its name to several mines. The East Pool mine, just south of the road, was the most enduring of these; originally known as Pool Old Bal, it was worked for copper until 1784 and its section was illustrated by Borlase (1758, Pl.XVIII). Restarted in 1835 under its new name, it turned to tin production, also yielding considerable amounts of arsenic and tungsten. Neighbouring setts were acquired, notably Wheal Agar, across the road to the north, in 1896, and the Tolgus group. The letters EPAL on an old chimney stack, for 'East Pool and Agar Limited', are clearly visible from the road; it is most unusual for a mine to be labelled. The mine survived a major underground collapse near the main shaft, in 1921, and also the economic slump of the 1920s. The New Tolgus shaft, sunk at this time, revealed interesting features of the mineralization but no economic orebodies. Although output gradually declined, government subsidy kept the mine in production for tin and tungsten throughout the Second World War until its closure in 1949.

Fig. 50 Dolcoath Copper mine, Camborne, Cornwall (T. Allom, 1831).

On the south side of the main road lies the still-active South Crofty mine (fig.32); restarted in 1854, in the period to 1905 it produced considerably more copper than tin, but the latter became increasingly important. After amalgamation with New Cook's Kitchen mine, followed by financial restructuring in 1905, new shafts were sunk and modern plant was installed. The 80-inch steam pumping engine at Robinson's shaft, which had earlier worked for 49 years until 1903 at Tregurtha Downs mine, was the last of the deep-mine Cornish beam engines to be replaced by electric pumps. It was a magnificent, unforgettable sight when in action, and took its last stroke in May 1955 after 101 years of hard work. The 90-inch engine at Taylor's shaft, East Pool, which ceased work a few years earlier, was one of the largest in Cornwall; until 1917 it had been draining the abandoned Carn Brea and Tincroft mines, at the Highburrow East shaft. These engines, and the 30-inch East Pool steam whim, are preserved and in the care of the National Trust.

None of the old mines south of Carn Brea has survived; originally copper producers, they changed to tin in the 1850s. To the south-west were the Grenville mines, which also became profitable for tin in the mid-nineteenth century, mainly on working the Great Flat Lode. These, and part of the adjoining South Condurrow mine, were amalgamated in 1912 to form Grenville United mines; they ceased work in 1920. King Edward mine is the renamed eastern section of South Condurrow, leased since 1897 by the Camborne School of Mines and used for practical training.

Between 1 and 2 miles to the east lay the most famous group of mines, centred on Carnkie; this contained the Basset mines (figs.52 and 53), which at depth worked the Great Flat Lode (amongst others). Ancient mines, they grew by incorporating adjoining setts and eventually

Fig. 51 Silhouette of the beam winding engine at East Pool mine, originally built in 1887 and now preserved in working order by the National Trust.

included South Wheal Frances and West Wheal Frances. Underground working ceased in 1919. There was little or no economic recovery of uranium, but good crystals of torbernite were found formerly; a cavernous black gossan also contained 'autunite', later shown to be bassetite (for the locality) and uranospathite (Hallimond, 1915). The 'uranite' (autunite or torbernite, they were not distinguished) from Wheal Basset and the nearby Wheal Buller is reported to have been very phosphorescent when first found (Garby, 1848:86).

An old miners' superstition held that the sighting of a 'Jack o'Lantern' indicated the position of a rich lode, and this was supposed to have been the case with North Wheal Basset around 1850. Digging where it had been seen by an old woman, the miners soon 'met with what has been described as one of the richest deposits of mala-

Fig. 52 Ruins of the Wheal Basset ore dressing floors.

Fig. 53 Ruins of the Wheal Basset ore dressing floors looking towards Carn Brea.

Fig. 54 South Crofty steam winding engine at Robinson's Shaft, made by Holman Bros., Camborne, 1907. Later replaced by an electrically driven machine (Photograph by J.C. Burrow).

Fig. 55 Incline skip road and ladder way of 450 foot level on the Great Flat Lode, King Edward Mine (South Condurrow), Camborne, Cornwall (Photograph by J.C. Burrow, early 1900s).

chite, red oxide, and black and grey copper ore ever found in the county. From it profits amounting to over £90,000 were made, and so magnificent was the mineral sent to surface that special precautions had to be taken against specimen hunters.' (Bluett, 1899:268). The old woman received a monthly payment, a new dress each year, and Grace's shaft was named after her.

South-east of Carn Brea, in the valley between it and Carn Marth, lay a further group of mines. Wheal Buller, to the east of Carnkie, was restarted in the 1850s and became one of Cornwall's important copper producers; containing little tin, it was abandoned in the copper slump of the late 1870s. Like several of its neighbours, it produced some good specimens of torbernite. To the north, immediately adjoining Carn Brea, lay Wheal Uny, which started before 1800 but ceased work in 1893; it produced tin, from the eastern end of the Great Flat Lode.

Fig. 56 The ore dressing floors of South Wheal Frances, Illogan. In the background is Bailey's or Engine Shaft of West Wheal Frances. They form part of the Basset mines on the south side of Carn Brea. The miners and bal-maidens are standing on and by the rag-frames, inclined tables over which the fine slimes or 'tailings' were washed from the central trough or launder. The heavier particles collected on the frames while the lighter material was washed away (Photograph by Bennetts of Camborne), *c.* 1890s).

Unusually for Cornwall, crystals of amethyst were found there.

3) Redruth - Gwennap - St Agnes (Fig.57)

This area includes Redruth, taking in the parishes of Gwennap, Chacewater, and Kea to the east, and of St Agnes and Perranzabuloe to the north. The parish of St Day was formerly part of Gwennap.

At the start of the seventeenth century, Uny-Redruth was a small parish 'wher are verie great store of Tynn workes, both Stream-workes and Load-workes' (Norden, 1728:42); unfortunately, its boundaries are rather uncertain. Actually in the town of Redruth, near the present railway station and alongside the old mineral railway, are the remains of the Pednandrea mine and, a little further on, of Wheal Sparnon. The water from these two mines was remarkably clean (Pryce, 1778:11). The Pednandrea lode produced mainly tin, with a little copper, while a parallel lode at Wheal Sparnon was rich in chalcopyrite. Rock crystal of cuttable quality occurred at Wheal Sparnon (Jars, 1781, 3:86). A cross-course – the Cobalt Lode – was worked for that metal around 1808 and later (Jenkin, 1962(III):20); it also contained a little gold. Pednandrea was the last to close, in 1891.

About a mile-and-a-half to the south-east of Redruth lies the summit of the Carn Marth granite mass. George

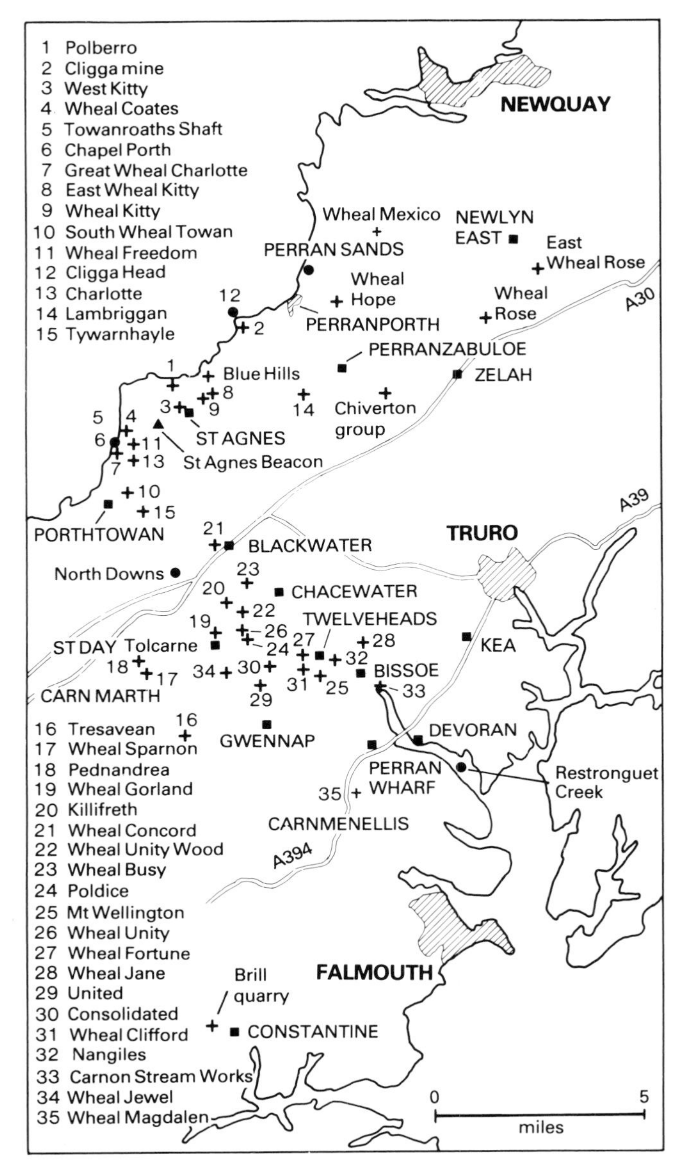

Fig. 57 Map of the Redruth – Gwennap – St Agnes area.

Henwood, in the late 1850s, estimated that minerals to the value of more than £30 000 000 had been raised within a radius of 2 miles of that spot (Burt, 1972:178). To the north and south of a line drawn from it to the village of Bissoe, about 4 miles to the east, is the famous mining parish of Gwennap, bounded on its north-east by the Carnon valley. This valley was formerly a rich source of stream tin, and the celebrated Carnon stream works lay further to the south-east, at the head of Restronguet Creek; extending over an area some 300 yds by a mile, the tin pebbles formed a bed from 4 to 6 ft thick at a depth of 36 ft below the surface.

Gwennap was the chief copper-producing parish in Cornwall, from the latter part of the eighteenth to the middle of the nineteenth century, with many large and important mines lying close to each other. Its reputation as a 'copper parish' has, however, overshadowed its earlier richness in tin: at the start of the seventeenth century, Norden recorded that it was 'A parishe riche of Tynn workes' (1728:45), with no mention of copper.

In the southern part of the parish, at the edge of the Carnmenellis granite, lay the Tresavean mine. First worked before 1740, by the 1830s it had become the third largest copper mine in the country and employed 1350 persons: between 1830 and 1837 it produced 9 per cent of Cornwall's total copper output. Already the site of experiments to make ladders easier for the miners to climb (Henwood, 1843:89*), the enterprise of the adventurers made it the first of the Cornish mines to use a man-engine: a pilot model in 1841, to 26 fathoms only, was driven by water power. In the following year a steam-powered version worked to 100 fathoms, by 1845 to 248 fathoms, and had reached the full depth of 290 fathoms when the mine closed around 1858 (Tew, 1981:48). After yielding immense profits for copper, tin was found at depth and the mine was worked well into the present century.

At Pengreep, some 2 miles to the east of Tresavean and near a greenstone contact, a remarkable occurrence of 'cobalt ore' and native bismuth was found in 1754 (Borlase, 1758:129,130). It was soon exhausted, and further search was abandoned due to 'a prodigious influx of water' (Pryce, 1778:50).

About a mile further north, to the east of Carharrack, lay the United group of mines, with the Consolidated group of mines another half-mile north again. These groups were formed around 1820 (James, 1949:196), or earlier in the case of the Consolidated Mines (Warner, 1809:290), and brought together a number of older, smaller mines: Poldory, and the Ale and Cakes mine, were the principal members of the United group; while Wheal Virgin, East and West Wheal Virgin, Wheal Fortune, and the Carharrack mine belonged to the Consolidated group. The lode systems of the two groups lay almost parallel to each other, those of the United group extending further to the west and those of the Consolidated group further to the east.

Individual mines of the United group were originally worked for tin, and when they changed over to copper, about 1750, they were known as the Metal works (Borlase, 1758:206; James, 1949:199). Worked at a loss between 1820 and 1840, they returned to prosperity and at its peak the group employed 1300 persons; the mines were very hot at depth, and temperatures of 114 °F were recorded in the 'Hot Lode'. The Ale & Cakes mine, at 285 fathoms, was the deepest of the United group.

Wheal Virgin was remarkably rich from the outcrop of its lode; in the first two weeks of working, in 1757, it produced ore worth £5700 at a cost of £100 and, in the next 23 days, a value of £9600 at a proportional cost of little more (Borlase 1758:206). It produced much native

Fig. 58 Sunset over Wheal Coates, St. Agnes Beacon, Cornwall.

copper, in pieces weighing up to 20 or 30 pounds (Klaproth, 1787:25), and by 1794 it was down to 160 fathoms (Maton, 1797, 1:244). Native copper was known as 'virgin copper', which may perhaps account for the mine's name. The Carharrack mine produced the first-recognized specimens of olivenite and of pharmacosiderite (Klaproth, *ibid.*:29; Bournon, 1801).

From 1819-1840, while the United Mines were doing badly, the Consolidated Mines were the largest and richest of all copper mines in Cornwall (Henwood 1843:88*); between 1823 and 1835 they accounted for over 15 per cent of the county's production, and in the late 1830s employed over 2000 persons.

Road transport for more than a few miles was difficult, and in the early days the Gwennap mines were served by the tiny port of Bissoe (Francis, 1845:7). With the building of the first part of the Redruth and Chasewater railway, opened in 1826, horse-drawn trucks with a capacity of 2.5 tons made the journey to and from the tidal port of Devoran, on Restronguet Creek, taking the ore for shipment to South Wales (Barton, 1960). They also carried supplies to the mines: Welsh coal for the pumping and winding engines (16 in all); Russian pitch, hemp, leather (for buckets and pump valves), and tallow (for candles); Norwegian support timber; Polish timber for pump rods; and so forth (Francis, 1845:62). Now that dredging has long ceased, and the mudflats are undisturbed by man, it is difficult to imagine Devoran as it was, until 1838, a more important port than Hayle. Opposite Devoran, at Perran Wharf, there were engineering works and a foundry.

Fortunes were made, but the richness of the lodes began to decline and in 1857 the mines were abandoned. The setts were taken over by the still-productive United group, and in 1861 the two groups combined with Wheal Clifford and other mines on their eastern side to form the Clifford Amalgamated Mines. All were abandoned in 1870.

A valley, carrying a branch of the Carnon stream, formed the northern boundary of the Consolidated Mines; and across this lies the St Day area. The St Day United Mines were formed in 1864, by the amalgamation of several much older mines; at this time, the group employed 553 people, and had nine (or more) steam engines at work (Collins, 1912:555).

Most of the mines in the St Day United group, although rich in copper, carried little tin at depth and so were forced to close in the copper depression of the later nineteenth century. This seems also to have been true of the ancient Poldice mine, already at work in the early 1600s (Norden, 1728:52, 'Poldeese'); it was an uncommonly rich tin mine for most of the eighteenth century and, on becoming a less-rich copper mine, produced little tin thereafter. By around 1720 it was, at 70 fathoms, the deepest of all the Cornish tin mines (Woodward, 1729:199,202); and before 1737 it was down to 106 fathoms, having produced tin to the value of £500,000 and employed 1200 workmen (Hals, in Borlase 1865:27). The County Adit was started in 1748, by the manager, John Williams, to reduce the cost of its drainage. This John Williams (1685-1761), incidentally, was the founder of the fortunes of the famous Williams family, who in the nineteenth century owned most of the Gwennap mines. It had largely turned over to copper production, apparently on an adjacent lode, and had reached 130 fathoms by 1791, the year in which E. D. Clarke (later professor of mineralogy at Cambridge) descended the mine; his account of the experience is vividly graphic (1793:90-98; Jenkin 1948:102). The eventual depth was about 250 fathoms. The Poldice copper ore was mainly chalcopyrite, but there was some crystallized tetrahedrite (Maton, 1797, 1:245).

The St Day mines, and particularly Wheal Gorland, Wheal Unity, and Wheal Unity Wood, are of special interest to mineral collectors for their olivenite ('wood copper') and other remarkably fine secondary copper minerals. Although the exact locality is not known, it is probable that the original specimens of cornwallite (with olivenite) were from this area. Wheal Gorland was 'so rich at times that the miners have been placed under a strict surveillance lest the Adventurers should be defrauded of the valuable minerals and rich ores' (Francis, 1845:11, footnote). The oxidized zone of the Muttrell Lode, first worked in the 1790s, was the richest source; it yielded, amongst many other species, excellent octahedral crystals of cuprite, rounded clinoclase groups, blue (and green) crystals of liroconite, and the first well-described specimens of the iron arsenate pharmacosiderite (Phillips, 1811). It should be noted that Phillips' detailed, contemporary account of the Great Gossan and Muttrell lodes is not readily comparable with later ones (Dines, 1956, 1:405). Wheal Gorland was abandoned before 1864, but a brief reworking of the upper levels until 1909 yielded a little tin and some wolframite. Apart from copper ores, Wheal Gorland also produced considerable amounts of fluorite. Its dumps, which for many years had provided happy hunting for collectors, were carted away for processing in the 1970s; during this final removal, many specimens of ceruléite were found.

In addition to most of the species found in Wheal Gorland, Wheal Unity produced fine brown prisms of mimetite, some of them hollow, and on these the presence of chloride in the composition was first established (Gregor 1809). The Tolcarne mine, mined for tin in the early seventeenth century (Norden, 1728:52, 'Tolkerne'), lay about half-a-mile south-west of Wheal Gorland, and produced torbernite of specimen quality (Sowerby, 1817, 5:Tab.CCCCLXXXVII).

On the hillside above the western bank of the Carnon valley, mid-way between Twelveheads and Bissoe and near the mouth of the County Adit, lies the Mount Wellington mine. Recently closed, it first started in the 1920s and was reopened in the 1970s to work in conjunction with Wheal Jane. Mount Wellington, and the old Ale and Cakes nearby, are the only two mines in the western half of Cornwall to have produced specimens of crystallized baryte.

Wheal Jane, first worked in the eighteenth century and still active, lies in the adjacent parish of Kea, about a mile-and-a-half east of Mount Wellington mine across the Carnon valley. The earliest principal output of Wheal Jane was pyrite, used for the manufacture of sulphuric acid, but complex lodes are now being worked for tin. Between the two mines, and working the same lodes, was the Nangiles mine, where copper, tin, zinc, and iron ores were all worked. Wheal Jane is the type locality for ludlamite (1877), in cavities of the pyrite in association with vivianite, and well-crystallized specimens have been found recently; unfortunately, the pyrite matrix is very susceptible to decomposition in storage.

Just outside Gwennap parish, near Chacewater and north of the road from Truro to Redruth, lay Wheal Busy; nearby, and south of the road, lay the Killifreth mine, which was worked for tin in the late 1800s. Wheal Busy, earlier known as the Chacewater mine, is thought to have started work before 1720, and it headed Borlase's list of 'the first and greatest' copper mines (1758:206). Production seems to have changed to tin in the middle years of the nineteenth century, followed by a considerable quantity of arsenopyrite around 1900. A Newcomen pumping engine was installed in 1727, which was followed in 1775 by a 72-inch model, the largest in Cornwall during the eighteenth century. This engine was rebuilt three years later by James Watt, to incorporate his patented condenser; but the first of Watt's Cornish engines was also built here, a few months earlier. The last engine to be installed at Wheal Busy was of 85 inches diameter, in 1909, shortly before the mine finally ceased work.

About a mile to the west of Wheal Busy, in the North Downs area, is a small and loosely-defined group of mines. The most northerly, Wheal Concord, near Blackwater, is a small and very old mine which has been worked at various times since the 1700s; recently reopened, in the early 1980s, it is still at work in 1986 (Congdon, 1982).

To the north of Redruth, on the coast, Portreath was once a busy port serving many of the Camborne-Redruth mines (Hitchins & Drew, 1824, 2:333). A little to the east, just beyond Porthtowan and at the western edge of St Agnes parish, lay several rich copper mines. These included the Towan group, United Hills, and the Tywarnhayle mine; Tywarnhayle was the first of the Cornish mines to be equipped with electrically-driven centrifugal pumps (Thomas, 1907). South Towan was one of the mines at which chalcopyrite was regarded as worthless when originally found (Francis, 1845:8, footnote). More intensively worked in the first half of the nineteenth century, they produced such fine specimens of crystallized chalcopyrite that Brooke & Miller, in their *Mineralogy* (1852), gave it the name 'towanite' (these authors also favoured the name 'redruthite', of Nicol (1849), for chalcosine). Great Wheal Charlotte, just south of Chapel Porth, worked a copper lode with some levels extending a short distance beneath the sea. Inland lay Wheal Freedom and North Wheal Towan, later renamed Charlotte United; the old engine house is now owned by the National Trust, as also is that of Wheal Coates to the north of Chapel Porth.

Wheal Coates, on the western side of St Agnes Beacon, was worked for both copper and tin during the last century but was never particularly rich; the Towanroath engine shaft, at the cliff edge, is 106 fathoms deep. The mine is well known to mineral collectors for its fine pseudomorphs of cassiterite after sanidine-habit orthoclase, and svanbergite has been reported from the andalusite schist there (Kingsbury, 1964:249). Pits dug nearby, on the slopes of St Agnes Beacon, were once the main source of 'candle clay', a sticky pipeclay used by miners all over Cornwall to secure candles to their hard hats (Hogg, 1825:171; Hawkins, 1832:138; Macfadyen, 1970:29).

Around St Agnes there are many old and rich tin mines, the most easterly of the areas mentioned by Norden (1728:52,65). On the north side of St Agnes Beacon ('St Anns Ball'), on the cliffs, lay the Polberro mine:

> 'Great part of the ore [from Polberro] was so rich and pure that it needed not to be stamped, and the lode is so large that it affords vast rocks of tin: one rock, in March 1750, was brought to Killinick [Calenick] melting-house near Truro, which . . . weighed six hundred and sixty-four pounds . . . [and another] weighed 1200 pounds.' (Borlase, 1758:188).

A great many old setts were later incorporated in the Polberro group, including Wheal Trevaunance and the Pell mine which yielded superb specimens of fluorite, with crystals showing the 'four-faced cube' form {310} to perfection. On the eastern slopes of St Agnes Beacon, between St Agnes and the sea, West Wheal Kitty was a group of some 30 older setts; one of these was Wheal Rock, where the first stannite specimens were found by Raspe in 1785. The multiplicity of ownership of the setts made working difficult. West Wheal Kitty, with workings under St Agnes, closed in 1919; it was reopened in 1926, and a new rich tin lode was worked, but it closed again in the world slump of 1930. Many old specimens, showing superb sphenoidal crystals of chalcopyrite, are simply labelled 'St Agnes'; they probably came from mines of the West Kitty group, including New Wheal Kitty. It is possible, however, that the bare 'St Agnes' locality might also have encompassed material from the Towan mines;

and it should be noted that there seem to have been two mines called Wheal Rock in St Agnes parish (Collins, 1912:517,568; Jenkin, 1962(II):15,28).

At some time before 1818 another mine in the West Kitty group, Wheal Kind (or Kine), began to produce the best crystals of vivianite known to that date, up to 2 inches long, and sales of the specimens were held at the mine (Jenkin, 1962(II):26). It may well be that this mine, specimens from which are usually labelled simply 'St Agnes', is the type locality for vivianite. A Wheal Kine specimen in the dealer Henry Heuland's possession was figured as 'phosphate of iron' by Sowerby, with the information that it was found in a north-to-south vein at a depth of 55 fathoms (1817, 5:Tab.DXLIV and p.270).

To the east of St Agnes across the valley, Trevaunance Coombe, lay Wheal Kitty and the various mines of the Blue Hills group. The richer parts of the old dumps of all these mines have been re-worked at various times, and considerable quantities of black tin have been recovered from the beach sands of Trevaunance Cove; pebbles of 'wood tin' have also been found here.

Further to the north-east of St Agnes were numerous old workings, for both copper and tin, and on the coast are the famous cliff lodes and greisen mineralization at Cligga Head. The original specimens of 'isostannite' (now ferrokësterite) came from sulphide segregations at Cligga, and many specimens from the adits at beach level have reached the market. North of Perranporth, and amid the dunes of Perran Sands, parts of the open-trench workings of the Perran Iron Lode can still be seen. This massive lode was worked for both iron and zinc ores, and has been traced inland some 4 miles. For a distance of 2 or 3 miles to the south of Perranporth, along the valley, there were many small lead and zinc mines; Lambriggan produced some unusually pale-coloured sphalerite. In this area, too, was Wheal Mexico, near Perranzabuloe, a visit

Fig. 59 Minerals for sale at Wheal Kind, St Agnes, Cornwall. Advertisement in the Royal Cornwall Gazette, September 26, 1818.

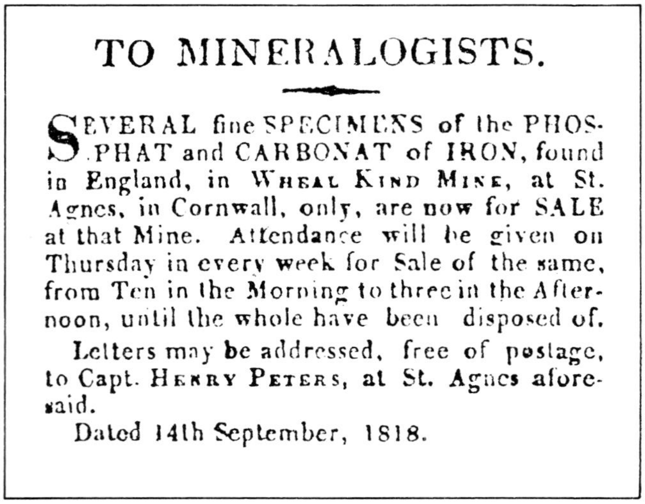

TO MINERALOGISTS.

SEVERAL fine SPECIMENS of the PHOSPHAT and CARBONAT of IRON, found in England, in WHEAL KIND MINE, at St. Agnes, in Cornwall, only, are now for SALE at that Mine. Attendance will be given on Thursday in every week for Sale of the same, from Ten in the Morning to three in the Afternoon, until the whole have been disposed of.

Letters may be addressed, free of postage, to Capt. HENRY PETERS, at St. Agnes aforesaid.

Dated 14th September, 1818.

to which was briefly described by Maton (1797, 1:252). A small mine, it struck silver-rich ore in 1785 which yielded some very pretty crystals of chlorargyrite (Sowerby, 1809, 3:Tab CCXLIV). Pyromorphite prisms, more or less completely pseudomorphed by galena, were found at Wheal Hope, also nearby.

The principal lead-mining district of Cornwall, however, lay further inland. The Chiverton group of mines, north of Truro and near to the village of Zelah, produced moderate amounts of lead and copper; but their output was modest compared with those of Wheal Rose and especially East Wheal Rose, some 2 miles away near the village of Newlyn East. In the second quarter of the nineteenth century, East Wheal Rose was by far the largest lead producer in Cornwall (Douch, 1964).

4) Central and North Cornwall: St Austell - St Endellion - Delabole - Liskeard (Fig.61)

A few miles to the east of Truro lies the granite of St Austell Moor, with huge conical heaps – now truncated – of waste quartz and mica from the china clay pits forming an impressive and characteristic artificial landscape over much of it. In early times, tin streaming was extensively practised on the St Austell moorlands; the once-important Poth [Porth] streamworks lay near Par, between St Austell and Fowey (Maton, 1797, 1:152; section in Rashleigh, 1802), and were exploited underwater using an iron caisson (Hitchins & Drew, 1824, 2:67). To the east of St Austell lies the estuary of the Fowey, the upper reaches of which drain Bodmin Moor. These areas formed much of the ancient stannary districts of Blackmore and Foweymore, respectively. From the start of the nineteenth century, there were relatively few large or deep mines in the area, and metalliferous mining tended to diminish with the rapid expansion of the china clay industry; minor mineralization, however, is constantly being exposed in the working clay pits. In the days of steam power, beam pumping engines were commonly used in the clay pits, and the remains of many of the old engine houses may still be seen in the district. The first attempts to smelt copper in Cornwall were made on the Polruddon estate, in St Austell parish (Hitchins & Drew, 1824, 2:58).

The earliest workings for china clay were toward the western part of the St Austell granite, in the parish of St Stephen-in-Brannel; and among these, south of St Dennis, was the old Stennagwyn mine, worked for both copper and tin and which also produced some stannite. The Rev. William Gregor, vicar at the nearby parish of Creed, analysed various minerals from Stennagwyn including wavellite and autunite / torbernite (Gregor, 1805); he also supplied some specimens from this mine to Sowerby, for inclusion in his *British Mineralogy*, among them the first known crystals of fluellite (1809, 3:Tab.CCXLIII).

In the Fal valley, outside the granite and some two

Fig. 60 Carclaze opencast tin mine, near St Austell (T. Allom, 1831).

miles south-west of Stennagwyn, was the South Terras mine. Small quantities of torbernite and autunite had long been known from the area, but a fairly rich vein was found about 1880 (?) and the South Terras mine became a leading producer of uranium for three or four decades (Collins, 1912:241). It is also notable as the locality for neotype specimens of the rare nickel arsenates, xanthiosite and aerugite (Davis, Hey, & Kingsbury, 1965). St Austell Consols, a tin mine about 2 miles to the east, also yielded commercial quantities of uranium ore and produced small amounts of nickel.

Although there were few tin mines on the southern edge of the granite, the Polgooth mine, some 2 miles south-west of St Austell, more than made up for the lack in numbers; St Austell may have owed its growth, from village to town, to the commerce induced by the mine (Hitchins & Drew, 1824, 2:47). In the early 1700s, it was

> 'The mine which has turned out the most gain, and the greatest quantity of tin as yet known . . . the adventurers have got twenty thousand pounds annually for a great number of years following.' (Borlase, 1758:189).

By 1794, there were

> 'no less than fifty shafts in Polgooth; twenty-six are still in use, with as many horizontal wheels, or whims. The main vein of ore, which is about six feet thick, runs from east to west . . . and there is another that cuts the former nearly at right angles, and consequently runs north and south . . . we were informed that [the mine] has afforded tin the full length of a mile.' (Maton, 1797, 1:155).

Closed for a few years, from 1806 to 1814, Polgooth remained active thereafter for most of the nineteenth century.

One of the oldest tin excavations in the area was the Carclaze open pit, just over a mile north of St Austell; a stockwork, with a short canal to transport its produce, it was described (without name) by Jars on his visit in 1764/5 (1781, 3:190). During much of the nineteenth century, it was worked for china clay.

By far the largest number of mines lay to the east of St Austell where, inland from the small ports of Charlestown and Par, they worked setts of east-to-west copper- and tin-bearing lodes, together with some lead and zinc. The largest tin producers were Eliza Consols, and the Charlestown and Cuddra mines; the Bucklers section of Charlestown produced small but sharp crystals of siderite. The major copper mines were Fowey Consols, Par Consols, and the Pembroke group. Fowey Consols, 1 mile north-east of St Blazey, is well-known for its fine specimens of bismuthinite (many encrusted with minute spherules of wood-tin), chalcotrichite, and rhabdophane; well-crystallized tetrahedrite came from the Crinnis mine, some 3 miles to the south west.

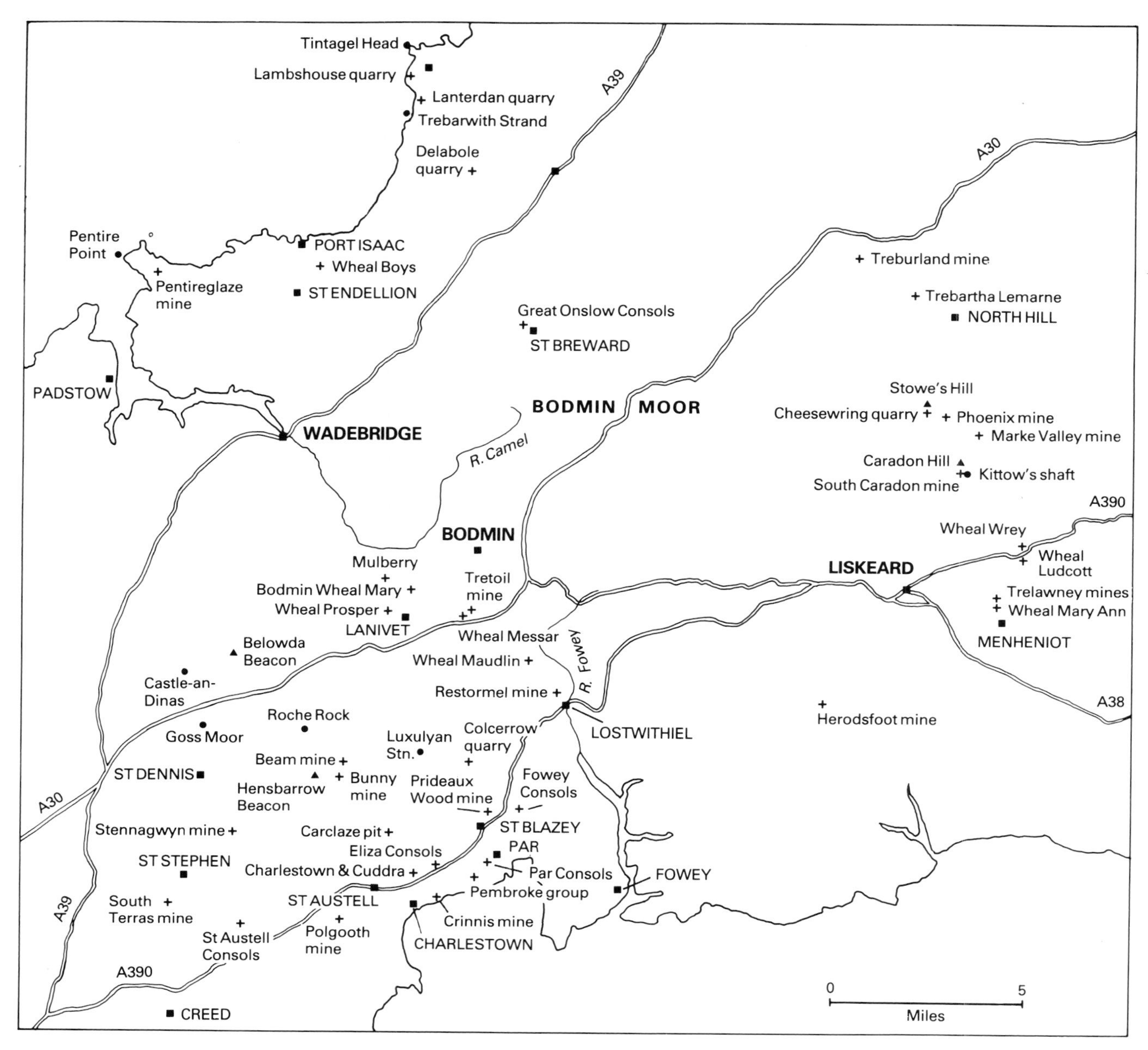

Fig. 61 Map of Central and North Cornwall: St Austell – St Endellion – Delabole – Liskeard area.

To the west of Fowey Consols, the small Prideaux Wood copper mine produced some curiously-fragmented wood-tin spherules in a granitic matrix. Further north again, near Luxulyan station, the Colcerrow quarry produced some small but perfect blue apatite crystals, in veins of a granite-pegmatite.

On the northern side of the St Austell granite, on Goss Moor to the north of St Dennis, lay one of the largest alluvial tin areas of the county, where both stream- and shallow surface-workings survived into the present century. Nearby is Roche Rock, a small outcrop of quartz–tourmaline. Two miles to the east of St Dennis, on the slopes of Hensbarrow Beacon, lie the dumps of the old Bunny mine from which the turquoise-group mineral rashleighite was first described. The Beam mine, a little to the north, yielded one of the few specimens of beryl known from Cornwall (Hallimond, 1939).

To the north of the main (A30) road are the two small rounded bosses of Castle-an-Dinas, topped by ancient earthwork fortifications, and Belowda Beacon. The Castle-an-Dinas mine, worked sporadically between 1918 and 1958 for wolframite, was small but rich; the vein ran in a north–south direction, and was also unusual in containing large amounts of löllingite at depth. It is the type locality for russellite, which was never found on matrix but only in the wolframite concentrates. Massive topaz was found in small trial workings near the summit of Belowda Beacon.

To the north-east of Belowda Beacon, the chief tin producers in the area were the stockworks of Mulberry and

Wheal Prosper, both worked as open pits. The dumps of a small mine to the east of Wheal Prosper, Bodmin Wheal Mary, yielded some yellow crystals of helvine (Kingsbury, 1961:927). Again to the east, lying just to the north of the new Bodmin by-pass, are the remains of the Tretoil mine, which was probably the type locality for churchite (Kingsbury, 1956), and of the Wheal Messar dumps which yielded the only known Cornish specimens of ilvaite.

Between the St Austell and the Bodmin Moor granite masses, the north-to-south cross-course of the Restormel mine was a major producer of iron ore during the last century. The mine was known as 'Royal' after it was visited by Queen Victoria in 1846. Collins (1912:247) counted 36 alternating bands of quartz and iron oxide in the vein, indicative of episodic mineralization. The quality of some of its goethite crystals is outstanding. A little to the north and west of Restormel, Wheal Maudlin worked essentially east-to-west lodes carrying tin and copper. It produced a variety of less-common minerals (*ibid.*:246, footnote), including scheelite and cronstedtite; and its concentrically-zoned siderite crystals, possibly from a single pocket, were valued highly.

Much further north along the coast, in the area between Padstow and Port Isaac, there were many lead- and antimony-rich lodes. The presence of antimony was known in the seventeenth century, and stibnite was recorded from an unknown site on the Manor of Roscarrock (Grew, 1681:334; Borlase, 1758:129). None of the mines was large, but some excellent crystallized mineral specimens were produced, such as the acicular cerussite from the Pentireglaze mine. The parish of St Endellion contained antimony deposits (Woodward, 1728:20); the first specimens of bournonite, figured by Rashleigh (1797, Pl.XIX) as 'ore of antimony', came from Wheal Boys, around 1789, and the name 'endellionite' was used for long afterwards. Most of the Wheal Boys bournonite crystals are on a matrix of massive jamesonite, with a little sphalerite.

Five miles to the north-east of Port Isaac, and about two miles from the coast, is the large and ancient quarry at Delabole (formerly, Denyball), renowned for the high quality of its roofing slate. Still active, and open to visitors, it was already large over two centuries ago (Borlase, 1758:93; Hockaday, in Page, 1906:519). Mineralized fissures in the slate have long been known to yield an occasional good specimen of crystallized quartz, with adularia. The much smaller Lambshouse quarry (near King Arthur's Castle, Tintagel Head), and the Lanterdan quarry (about a mile to the south, at the north end of Trebarwith Strand), have produced small crystals of monazite and anatase (Bowman, 1900).

The north-east edge of Bodmin Moor has relatively little of mineralogical interest, apart from the very small manganese mine at Treburland (Russell, 1946) and tin-tungsten mineralization at Trebartha Lemarne, near North Hill. The south-east part of the moor, however, had two rich groups of copper (and tin) mines: around Stowe's Hill, topped by a picturesque tor on the edge of the Cheesewring granite quarry, lay the Phoenix group; and the Caradon group lay in a belt immediately to the south of Caradon Hill. Many of the long-derelict engine houses form prominent landmarks.

Tin streaming was carried on in the area from the earliest times, and in the nineteenth century excavations at the Phoenix site exposed small heaps of ancient concentrates buried by 4 or 5 feet of peat. The 'backs' of Stowe's lode were first mined for tin in the early eighteenth century. Copper mining began in the 1830s, when a group of miners led by the Clymo family drove a trial level in a small valley to the south of Caradon Hill, and they cut a rich east-to-west copper lode just before their initial capital ran out; this became the South Caradon mine. As a mining curiosity, Kittow's Shaft, at the eastern end of the sett, was remarkable for having three distinct inclines running from it in different directions. Apart from some brief setbacks, one around 1850 when local landowners had to be compensated for damage caused by the operations, copper mining was carried out with considerable profit for about fifty years, and attracted large numbers of miners from the less-prosperous western areas of the county.

The Caradon finds led to exploration at the Phoenix and Marke Valley mines, north-west of Caradon Hill and overlooked by the Cheesewring granite quarries, and in the 1840s copper ores were discovered below the tin-rich gossans. After the copper ran out, these mines became the largest tin producers in the eastern part of Cornwall. The Phoenix United mines closed in 1898; a brief period of reopening in 1907, with a new (Prince of Wales) shaft and engine, was abandoned in 1914 when it was discovered that there were only small patches of good ore at depth. The Phoenix mines produced many exceptionally well-crystallized specimens of cuprite, and the oxidized zone was remarkable for its variety of phosphate species, including chalcosiderite, andrewsite, and very fine libethenite. Good specimens of connellite are known from the Marke Valley mine, which is also the type locality for liskeardite. Among the better specimens from the Caradon mines were some superb fluorites.

Well to the south of the Bodmin Moor granite, south-west and south-east of Liskeard, north-to-south lead-silver lodes were rediscovered in the 1840s; rich deposits were worked profitably at Herodsfoot, and around Menheniot. Although the total output was not large, several of the mines also became known as a source of fine mineral specimens, largely through the enterprise of the dealer Richard Talling who lived at Lostwithiel. The old Herodsfoot mine was reopened for argentiferous galena in 1844:

> 'Antimony [ore], in small quantities, merely hand-specimens, is also disseminated through the lode;

Fig. 62 Engine house above 'Prince of Wales Shaft', Phoenix mine, north west of Caradon Hill, Cornwall. In the background can be seen the Cheesewring granite quarry.

Fig. 63 Remains of the Herodsfoot mine, Lanreath, Cornwall.

this principally occurs just below the gossan, and fortunately [!!] decreases in the lower levels, for where it most plentifully occurs, the lode is deteriorated.' (Giles, 1852:202).

The principal antimony minerals were bournonite and tetrahedrite, and both occurred well-crystallized. The best crystals of the former are magnificent, and have set a standard for the species that has yet to be surpassed; they seem (from the evidence of Talling's invoices) to have been found between 1858 and 1868.

To the east of Liskeard, the lodes have been worked for more than 3 miles northwards from Menheniot. Wheal Mary Ann and the Trelawney mines were prolific producers of lead and silver; although Wheal Wrey and the Ludcott group were of lesser importance, Wheal Ludcott was the first recorded Cornish locality for stephanite (Davies, 1866). The best specimens from these mines were of gangue minerals. The relative scarcity of crystallized calcite in most of Cornwall has been remarked by many writers, but the pseudo-prismatic acute rhombohedra from Wheal Wrey are noteworthy by any standards. Wheal Mary Ann produced many greenish or yellowish fluorites in beautifully-zoned cubes, often with corners characteristically modified by hexoctahedral

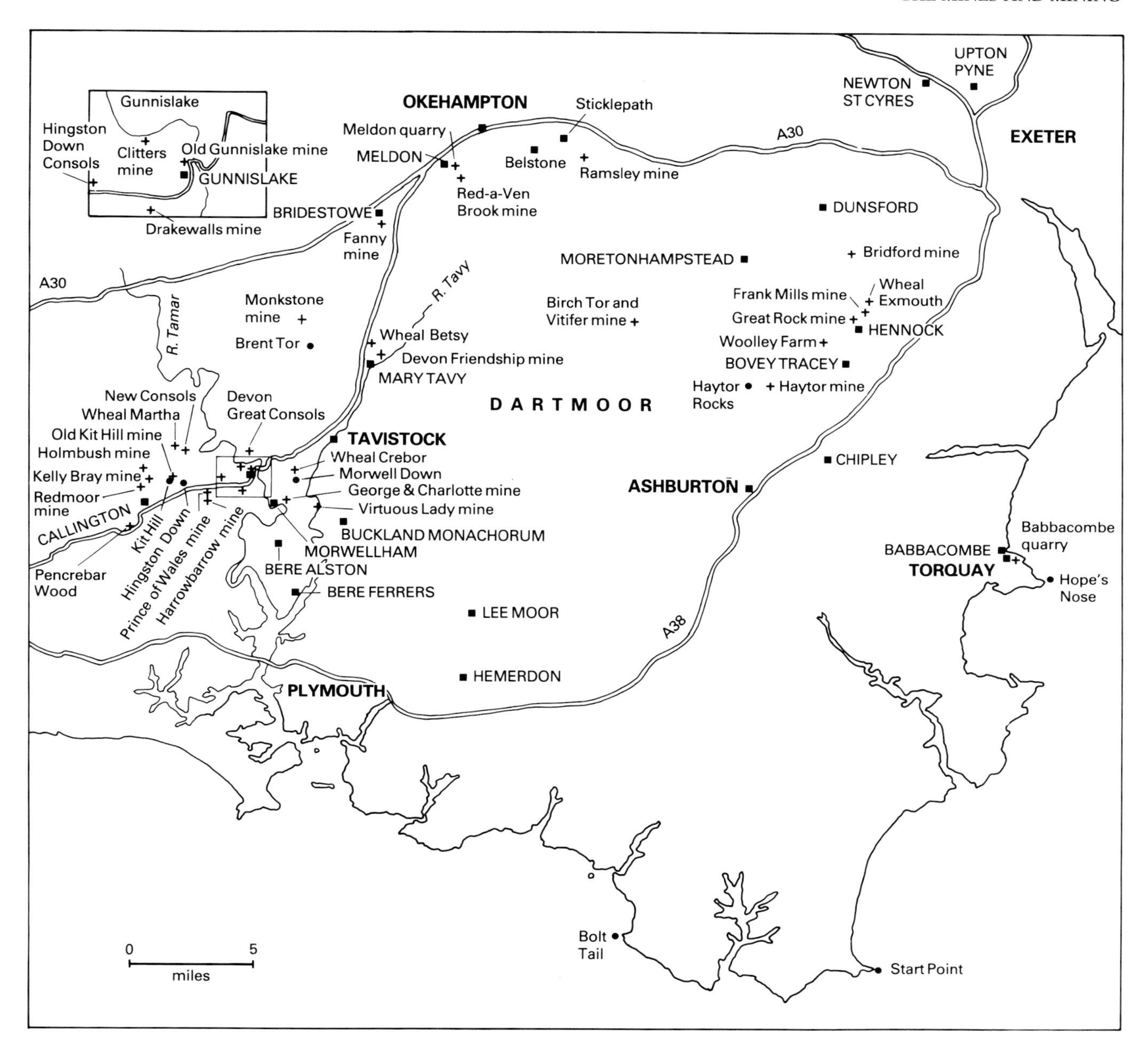

Fig. 64 Map of the East Cornwall – Devon area.

faces; also, some fine yellow-orange tabular baryte crystals.

5) East Cornwall - Devon (Fig.64)

The remaining Cornish tin and copper mines of any importance, and the largest of those in Devon, lay close to the granite ridge linking the outcropping masses of Bodmin Moor and Dartmoor. Most of this ridge, centred roughly on a line from Caradon Hill to Tavistock, is concealed; but it is expressed at the surface in the small masses of Kit Hill, topped by the chimney of the old Kit Hill mine, and Hingston Down. Immediately east of Hingston Down, at Gunnislake, it is cut by the Tamar valley. The river Tamar, which has its source north of Stratton, near the north coast, forms the boundary between Cornwall and Devon for most of its length.

Throughout this mineralized belt, the strike of the tin and copper lodes is east-to-west. They also yielded large quantities of pyrite and arsenopyrite, and wolframite was found to be widely associated with the tin-bearing zones. Streaming and shallow tin mining were carried on from the earliest times, and before the charter of 1305 the combined tinners of Cornwall and Devon used to meet in session on 'Hingston Hill' (Hunt, 1884:49). Later, the only one of the deeper mines in East Cornwall to yield more tin ore than other economic minerals seems to have been the Drakewalls mine, to the south of the Gunnislake granite. Stockworks yielding cassiterite and wolframite have been mined at Kit Hill and, in Devon, at Hemerdon. Lead-bearing cross-courses have also been worked, from ancient times in their own right and in the nineteenth century where they were intersected by copper-tin workings.

The Callington United mines, incorporated in 1888, included the much older Holmbush, Kelly Bray, and Redmoor mines, all of which had been productive. Situated west of the Kit Hill granite, they worked a series of east–west lodes for tin, copper, and arsenic, together with north–south lead-zinc cross-courses. Most of the group had closed by the end of the century, but parts of Redmoor were worked between 1907-1934. To the south of Callington, where the road passes through Pencrebor Wood, there is an abundance of loose lumps of rhodonite rock weathered black on the outside. The Kit Hill mine was worked in granite at the summit of the hill, and although never rich it produced some tin, tungsten, arsenic, and a little copper. In the valley to the north of Kit Hill, Wheal Martha was mined for copper in the 1860s and produced a little tin. In the belief that it lay near an 'emanative centre', and would be rich in tin at depth, it was re-opened as New Consols in 1946; unfortunately, the venture failed and the mine closed at the end of 1952. The Deerpark tunnel, earlier driven into the north side of Kit Hill, failed to intersect any worthwhile lodes.

On the south slopes of Kit Hill, the Harrowbarrow and Prince of Wales mines were small but rich in lead-silver ores. In this group of mines, Wheal Newton produced specimens of crystallized acanthite (after argentite) and stephanite that are excellent by any standards of comparison (Rudler, 1905:56).

East of Kit Hill, on the summit of Hingston Down, an engine house (built in 1903) still stands to mark the site of Hingston Down Consols; this commenced working for tin and copper about 1846, and was re-opened early in the present century in conjunction with Gunnislake Clitters mine and the Old Gunnislake mine. The Hingston Down Consols dumps yielded the type specimens of arthurite (Davis & Hey, 1964).

The Old Gunnislake mine, which had been started in the eighteenth century, produced the finest metatorbernite known from Cornwall at about the 90 fathom level (Phillips, 1816:115), and its sandy dumps behind Gunnislake village – although largely removed in the early 1970s – still yield occasional small specimens. Chalcophyllite also occurred at the Old Gunnislake mine (Hogg, 1825:35), and though the specimens were of inferior quality to those found at Wheal Gorland and other Gwennap mines it was named 'tamarite' by Brooke & Miller (1852), presumably for its proximity to the river; derivation of the name from Wheal Tamar, a smaller copper mine, seems less certain. Several beryllium minerals, notably bertrandite, have been found in small amounts in the area; the most recent of these is iron-rich roscherite, from the dumps of Gunnislake Clitters (Clark *et al.*, 1983). Most of the Gunnislake mines began working in the early 1800s, producing copper until the 1880s and arsenic from about 1865 to 1910; the nearby Drakewalls mine, mentioned above, was the only one to yield large quantities of tin.

On the other side of the Tamar lay by far the largest group of mines in this area, Devon Great Consols. Intensive mining commenced in 1844, although the gossans had been recognized many years earlier, and the main lode soon proved immensely rich; before operations ceased in 1901, it had been followed to a depth of 300 fathoms and traced horizontally for nearly 3 miles. The output of copper and arsenic was so considerable that the existing quay at Morwellham, on the Tamar, was considerably enlarged and linked to the mines by railway and an inclined plane of gradient 1:3 (fig.65). Copper ores were smelted in South Wales, whereas the calcination of the arsenical ores was carried out locally, and the resulting arsenious oxide was mostly exported to the United States to control boll weevil in the cotton fields.

The quay at Morwellham had several centuries of history before Devon Great Consols opened, and was the port of shipment for the stannary town of Tavistock from the Middle Ages (Booker, 1967). The river Tavy, for which Tavistock is named and which runs into the Tamar, was not navigable by barges; in the late 1700s, ores from the Devon Friendship copper mine, some 3 miles upstream from Tavistock, were laboriously hauled by road to Morwellham, over the high ground of Morwell Down between the Tavy and Tamar. The management of Devon Friendship was taken over in 1798 by the young John Taylor (1779-1863), later to become celebrated as a mining engineer, and one of his first projects was to improve transport by building the Tavistock Canal; the last 2 miles to Morwellham ran in a tunnel under Morwell Down, and took from 1803 to 1817 to drive (Taylor, 1817; Booker, 1967). Several lodes, none of them large, were intersected during the driving of the tunnel and later mined; and the mineral childrenite was first described on specimens taken from somewhere along its length, possibly Wheal Crebor. The George and Charlotte mine, close to Morwellham and first worked before 1815 (Lysons, 1822, 1:cclxxxiv), produced the largest known crystals of childrenite at some time before 1858; it has been re-opened as a tourist attraction in recent years, to illustrate nineteenth century mining, and in the course of this development further good specimens have been found. It is also the probable type locality for tavistockite, which has been discredited (Embrey & Fejer, 1969). Morwellham itself has been developed imaginatively as a museum of industrial archaeology.

Wheal Friendship, or Devon Friendship mine, near Mary Tavy, was already at work as a rich copper mine at the end of the eighteenth century; together with the adjoining mine, Wheal Betsy, it also produced much silver-rich galena. Its ores were highly arsenical, and the remains of the once-extensive flues and collecting

Fig. 65 Morwellham on the Tamar about 1870. Piles of ore may be seen on the quays. The paths of the ore transporting inclines for the Tavistock canal and from Devon Great Consols can be seen on the side of the hill in the background. The quays and other facilities at Morwellham have been restored as a working museum and the adjacent old George and Charlotte copper mine can be visited (Booker, 1967).

chambers of the roasting process may still be seen. Crystallized specimen minerals include calcite of 'schiefer spar' habit, scheelite, well-crystallized arsenopyrite, and pyrite; the curiously-curved crystals of the latter depicted by Sowerby (1817, 5: Tab CCCCLIV) were almost certainly from this mine (Maton, 1797, 1:296).

In the area between the Tamar and the Tavy, near Bere Alston and Bere Ferrers, the Tamar lead mines were among the earliest of all mining operations in South-West England and the galena in them was rich in silver. Long since abandoned, they produced fluorite crystals of complex habit in the early 1800s (Phillips, 1823); fluid-filled casts of quartz after fluorite, from the South Hooe mine, were once sold under the name 'water-cubes' (Hawkins, in Lysons, 1822, 1:cclxix).

Almost due south of Tavistock, in a beautiful position where the small river Walkham runs into the Tavy, lies the Virtuous Lady mine, Buckland Monachorum. This small copper mine was possibly first worked as early as the sixteenth century, on a shallow-dipping lode (in marked contrast to the steep dips of most in the area) varying in width from a few inches to 20 or 30 ft, and continued intermittently until about 1875. It is celebrated for the fine mineral specimens that have been preserved in collections, and its last period of working is said to have been mainly to obtain more of them (Jenkin, 1974:48). The best-known of these are siderite, forming two types of epimorphous cast from which the original mineral has been disssolved: the one, tapering 'lady's slippers', presumed to be after baryte crystals of unusual habit; the other in hollow boxes, after cubic crystals of fluorite. 'Capped quartz' is another curiosity, milky prisms in which the growth of the terminal faces was interrupted by the deposition of clay, so that – when broken – the ends come away as caps. Other crystallized species include pyrite, chalcopyrite, anatase, and large, wedge-shaped arsenopyrite (Russell, 1913). About 2 miles further to the east lay Wheal Franco, the type locality for francolite.

On the northern edge of Dartmoor, in the Okehampton area and within the metamorphic aureole of the granite, both tin and copper mineralization have been recorded from mines along the outcrop of the Meldon Formation (Lower Carboniferous) between Fanny mine, near Bridestowe, and Ramsley mine, Sticklepath (Dearman & Butcher 1959, map). The tin mineralization appears to be concentrated within skarns, and in narrow bands of wollastonite-hornfels containing tin-bearing garnets and – at the Red-a-Ven Brook mine – malayaite. Chalcopyrite and arsenopyrite occur in horizons of calc-silicate hornfelses that are rich in garnet and axinite, and were worked at the Belstone and Ramsley mines. Grossular, rarely well-crystallized, is abundant at Belstone; and at Ramsley, bipyramidal crystals of scheelite and hornstone

Fig. 66 Ivy clad ruins of one of the engine houses at the old Wheal Exmouth lead mine in the Teign Valley, Devon. As the engine houses had to be built within view of the estate house they were constructed in an imposing castellated style.

pseudomorphs after them have been found (Kingsbury, 1964:249). The large quarry at Meldon, which has been excavated for railway ballast since the end of the nineteenth century, is in sediments extensively altered by the Dartmoor granite; in the 1950s and 1960s, manganese-rich beds yielded huge amounts of rhodonite and bustamite. A nearby aplite body has also been quarried which, with with associated pegmatites, has also produced many interesting minerals (Kingsbury, in Macfadyen, 1970:61-63).

Although it is generally considered that stream workings on the granite of Dartmoor once produced most of the tin from South-West England, from the Dark Ages to about 1200, there were later very few mines of any consequence. It may reasonably be concluded that most of the lodes were removed by erosion. The most important and extensive of the Dartmoor tin mines were the Birch Tor and Vitifer mines, 6 miles south-west of Moretonhampstead on the north-east side of the moor, where extensive trenches of the old open works and leats for the water-wheels can still be seen. All underground working stopped in 1915, and all surface work in 1926.

Also on the eastern side of Dartmoor, not far from Bovey Tracey, the Great Rock mine at Hennock was worked for micaceous hematite, and has produced fine pyrite specimens. Magnetite deposits were formerly worked at Haytor, the mine for which the 'haytorite' pseudomorphs of chalcedonic quartz after datolite were named. Black prisms of schorlite, up to four inches across and labelled 'Chudleigh' or 'Bovey Tracey', are to be found in many old collections and were figured by Sowerby (1817, 5:Tab DXLVIII); associated with white apatite, they were found some time after 1810 in a pegmatitic cavity at Woolley Farm (Mawe, 1818). Between Dunsford and Hennock, in the middle Teign Valley, lodes in a wide zone parallel to the granite contact have been worked for argentiferous galena, sphalerite and baryte. The largest lead-silver producers were the Frank Mills mine and Wheal Exmouth (fig.66), and the Bridford mine was worked for baryte (Polkinghorne, 1952; Schmitz, 1980).

No account of the mineral localities of Devon would be complete without at least passing mention of the ancient lead mines of Combe Martin, on the north coast, and of the High Down quarry at Haddon, near South Molton, which is the type locality for wavellite (the synonym 'devonite' was of very short duration). Near Exeter, at Upton Pyne and Newton St Cyres, well-crystallized manganese oxides were once found in opencasts on bedded deposits. Finally, near Torquay, the small quarry at Babbacombe produced specimens of pink cockscomb baryte that are represented in many old collections. The most recent interesting mineral discovery in this area was of the palladium species mertieite and isomertieite, associated with the already-known dendritic gold at Hope's Nose (Clark & Criddle, 1982).

Fig. 67 Talling invoice, April 10th 1871, with specimen no. 3994 Cuprite.

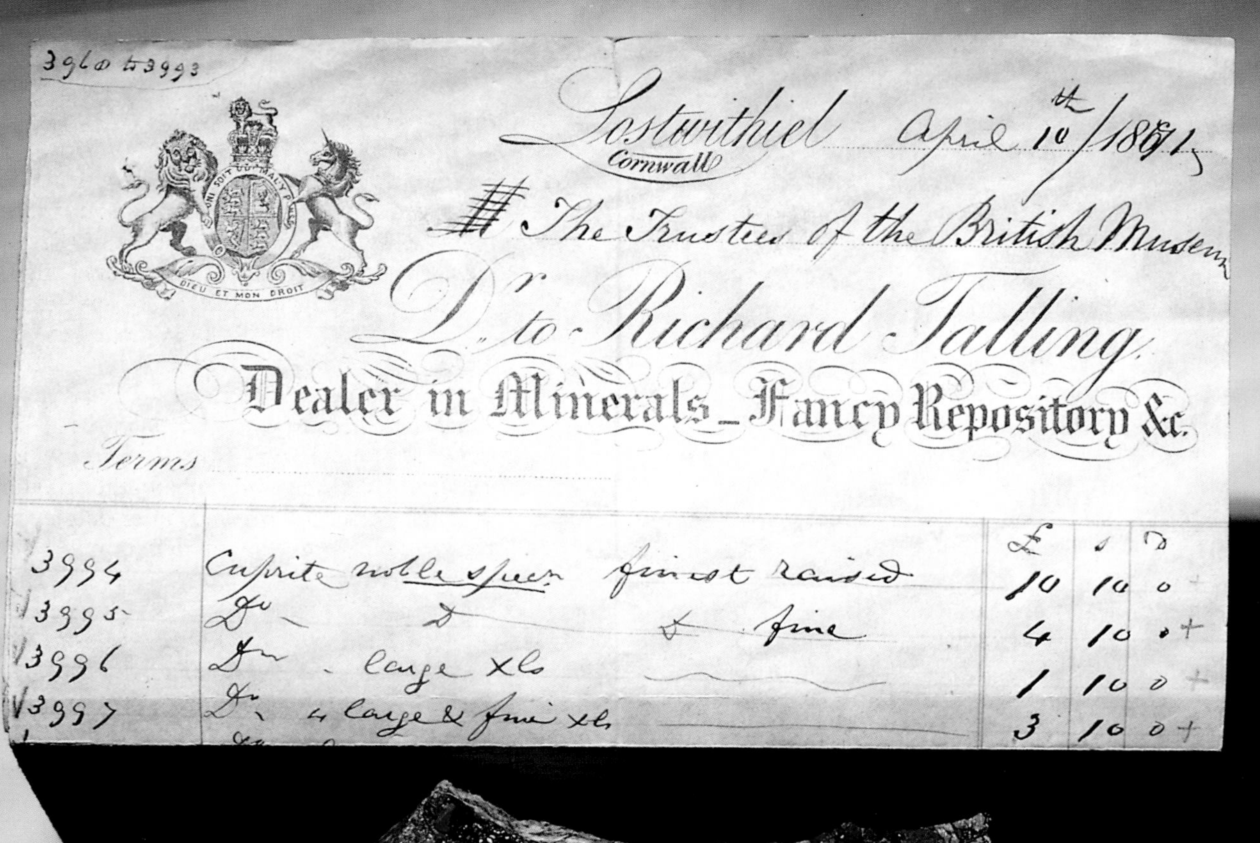

3960 to 3993

Lostwithiel
Cornwall
April 10th/18[illegible]

The Trustees of the British Museum

Dr. to Richard Talling

Dealer in Minerals_Fancy Repository &c.

Terms

		£	s	d
3994	Cuprite noble spec. finest raised	10	10	0 +
3995	Do. D. D. fine	4	10	0 +
3996	Do. large xls	1	10	0 +
3997	Do. 4 large & fine xls	3	10	0 +

COLLECTORS AND DEALERS

Specimens of Cornish minerals, more particularly of the many secondary copper species for which the county is rightly famed, are to be found in collections the world over. This section is a brief and highly selective account, intended to introduce to the reader a few of the more notable collectors and dealers whose activities have helped to bring about this distribution. Space considerations unfortunately preclude even passing mention of most of the collections outside the United Kingdom, however significant their content. Of the many institutional collections in the United Kingdom, attention has largely been confined to those of the Royal Institution of Cornwall, at the County Museum, Truro; the Royal Geological Society of Cornwall, Penzance; the Camborne School of Mines; the Plymouth City Museum; the British Museum (Natural History) and the Geological Museum, London; the University Museum, Oxford; and the Sedgwick Museum, Cambridge (most of the collections of the Department of Earth Sciences, Cambridge, are not on display).

Although Cornwall has been renowned for the economic value of its mineral produce for well over 2 millennia, its individual minerals have attracted interest for little more than three centuries; and its 'specimen mineralogy' is younger still. Dr Johann Joachim Becher (1635-?1682), who was invited to Cornwall in 1679 to introduce new methods of mining and smelting, published his *Alphabetum Minerale* in Truro in 1682; the work has an alchemical flavour, and contains almost no hard mineralogical information, but the dedication (to Robert Boyle) contains the words:

> 'I believe there is no place in the world which excels Cornwall in the quantity and variety of [its minerals]: and I confess I have found here a mining school, and from being a teacher am become a scholar.' (Klaproth, transl. Groschke, 1787).

Mineralogy as a science, in common with the other branches of natural history, owes an inestimable debt to the enthusiasm of amateur collectors; for it is nothing if not specimen-based, and our public museums would be poor things indeed had they not first been based on private collections. Much of their growth is due to generous gifts or bequests, and to the purchase of the whole or selected parts of private collections.

> 'How much it is to be wished that men of large fortunes and wide connections would more frequently thus cultivate some branch of useful knowledge . . . such men are ornaments to their country; and they are the best benefactors to science, for whilst they find an amusement for their own leisure they promote its popularity, and encourage its votaries.' (Maton, 1797, 1:150).

Even then, when there was far less distinction between the two than exists now, Maton might also have remarked how frequently the observant amateur has brought to the attention of the professional a new mineral, or an interesting feature or association of an old one.

Published information on mineral collectors is at best scanty, and it is not uncommon to find biographies, of people known to have had fine mineral collections, in which this facet of their activities is largely or wholly neglected. Even less is known of most of the dealers who attended to the collectors' needs. Without the private collector, the patronage of museums alone could never have supported the market, and there would have been little incentive for fine specimens to be saved from the jaws of the ore crusher. Specialist mineral dealers have always been rare, and dealers have commonly supplied all manner of other natural history specimens. As with the collectors, so with the dealers: minerals are seldom mentioned in lists of dealers in shells or fossils.

Dealers have usually had a wholly-different, principal source of livelihood, often of the kind most likely to have brought them into regular contact both with possible customers and with the main source of specimens, the working miner and his family. These trades have included printing and bookselling, or running a general store, an inn, or cobblers' or barbers' shops. Our knowledge is scantiest when it comes to the trading of stock between dealers; and the identities of the individual working miners or quarrymen who actually found the specimens are almost never known – not the least of the reasons being their near-certainty of dismissal or prosecution if found trading by the owner or manager, and the dealers' desire to conceal their sources from competitors.

General accounts of early collecting in geology and mineralogy, with particular reference to England, have been given by Campbell Smith (1978) and by Torrens

Fig. 68 'Minerals for Boots' – advertisement in *West Briton*, 19 March, 1841.

S. MICHELL, Mineralogist,

3, *LEMON-STREET, TRURO,*

BEGS to return thanks to the Nobility and Gentry of Cornwall, and the scientific gentlemen in general of Great Britain, for their liberal support during the last twenty years, and to announce that he has for Sale a large and superb collection of *genuine* Cornish and Foreign Minerals, among which are several great Novelties to the scientific world, two or three of the Cornish Specimens being quite new, and not to be found in any Cabinet in the county.

S. M. begs also to announce, that he has for Sale a superb collection of new and rare Shells, sent from the Philippine Islands by one of the greatest Conchologists of the age, and the first that has been brought from that locality to Enrope.

*** Boots and Shoes, of which S. M. has a large stock, given in exchange to Miners for good Minerals.

Dated, March 16th, 1841.

(1980). A series of five papers entitled 'An account of British mineral collectors and dealers in the 17th, 18th, and 19th centuries' were read before the Mineralogical Society between 1930 and 1934 by (Sir) Arthur Russell but, unfortunately, were not published as a whole; some appeared later (on Rashleigh, Hawkins, and Gregor), and fragmentary notes for some others have survived, information from which has been used below.

Before the beginning of the eighteenth century, specimens from Cornwall and the rest of south-west England featured hardly at all. Grew's catalogue (1681) of the collections of the Royal Society, in London, contained such entries as 'Bastard-Diamonds' [quartz], of which 'the Cornish are the best', and 'A pretty hard Ash-colour'd and Opacuous Spar [??], growing near the Tin-Mines'; these specimens, along with the rest of the minerals in the Royal Society's collection, were transferred to the British Museum in 1781 but have long been lost.

Dr John Woodward (1665-1728) was by far the most important of the early collectors, and the standards he set for documentation were wholly uncharacteristic of his time.

> '[His] museum no doubt resembled superficially other miscellaneous collections of the period, but it is to his credit that his collections had been amassed with the intention of advancing knowledge, and not merely to exhibit to friends and casual visitors.' (Eyles, 1971).

Woodward was 19 or 20 when he saw his first organic fossils; and, with the aim of explaining the natural history of the earth, he devoted the rest of his life to observing and collecting, both in person and through agents. The collecting rivalry and personal animosity that existed between Woodward and his great contemporary, Dr (later Sir) Hans Sloane (1660-1753), are well described by Levine (1977). Sloane's collections were, on the whole, superb, and occasioned the foundation of the British Museum in 1753; yet his minerals were poorly selected, and greatly inferior in both scientific interest and documentation to Woodward's.

The catalogue of Woodward's collection (1728, 1729), part of it published after his death, contained a wealth of meticulous detail, together with much general observation and speculation. The section on tin ores (Tome I, p.198-206) contains much factual detail, mostly provided by T. Lower, his agent. Thanks to the instructions for its care that he left in his will, together with the money to endow the Woodwardian Chair of Geology, the collection itself has been preserved in a remarkable state of completeness, in the Sedgwick Museum at Cambridge University. It is surprising, however, that his individual specimens have yet to be studied with attention comparable to that accorded to his life and theories; many are from Cornish localities, of which the spelling makes some look unfamiliar, and there can be no doubt that careful study would prove rewarding. Henckel's *Pyritologia*, unfortunately, contains nothing about Cornish specimens, because

> '. . . the English samples, dispatched for me by the celebrated Dr. Woodward, never came to hand.' (transl., 1758:2).

The Rev. William Borlase (1695-1772) was the first Cornishman to have formed a serious geological collection, and half of his *Natural History of Cornwall* (1758) is devoted to various aspects of the earth sciences and mining. In the year that this was published, he sent to London by the 'Turnpenny Tinship', for forwarding to the Ashmolean Museum at Oxford,

> 'all the Fossils of Cornish growth describ'd in my Nat: Histy . . . and . . . all the best specimens of Metals wch have fallen in my reach.' (letter, in Gunther, 1925, 3:376).

Borlase's illustrated catalogue, dated 1759, is in the Ashmolean; but

> 'Not a single specimen of the Borlase Geological collections placed in the care of Ashmole's keepers has survived. Borlase might as well have sent his treasures to Twickenham to add to the decorations of his friend Pope's grotto.' (ibid., p.223).

Some of the earliest surviving Cornish minerals at Oxford, now in the University Museum, were given and bequeathed much later, by Dr Richard Simmons (?1781-1846); his pictures were left to the National Gallery.

Borlase also gave mineral specimens to Leyden University, in exchange for plants (Pool, 1966:131). Some supposed native tin, that he gave to the Royal Society (Borlase, 1767), has not survived; and the only specimens now in Britain that are attributable to him seem to be a few that he gave to Thomas Pennant (1726-1798). Pennant's collection and catalogues are now in the British Museum (Natural History), but contain little else from Cornwall (Campbell Smith, 1913). The celebrated grotto of the poet Alexander Pope (Searle, 1745; see Mack, 1969), for which Borlase contributed quartz and other Cornish specimens, still survives in a rather sorry state; but the minerals have long gone.

Grottos, or artificial caverns (not necessarily underground) lined with sea shells and sparkling crystals, were a feature of many Italian gardens in the sixteenth and seventeenth centuries, forming decorative outdoor extensions of the indoor cabinets of natural curiosities. Pope's grotto at Twickenham was thus part of 'an older and outmoded tradition' (Hunt, 1985), but one which lingered on in Cornwall.

> 'the curious Mrs. Grace Percival of Pendarves . . . has offered us a fair pattern, by fixing side by side in her Fossilary an infinite number of crystals of

various and the clearest waters, in all shapes, single and in clusters, mostly out of mines in her own lands, all out of her neighbourhood. So many rich subjects will well remunerate the attentive inspection of every inquisitive Fossilist at her seat of Pendarves, in the Parish of Camborn' (Borlase, 1758:122).

This 'Fossilary', or grotto, must have been garishly spectacular when new, provoking an extravagant tribute in verse (Moore, 1755); by the present century it had fallen into ruin, and was finally demolished during the 1960s. Philip Rashleigh (see below), in the late eighteenth century and with the aid of a single tradesman (Gilbert, 1820, 2:875), built an octagonal grotto at Menabilly, near the shore, of which a contemporary picture has survived. An alcove lined with mineral specimens, similar in concept and probably dating from the same time, was uncovered when ivy was recently (1985) stripped from a garden wall at Clowance, once the home of Sir John St Aubyn (C. V. Smale, *priv. comm.*)

Mrs Grace Percival (1696-1763) may not have started a separate mineral collection, in addition to her fossilary; but by the nineteenth century there was a mineral cabinet at Pendarves (Gilbert, 1820, 2:695). On her death, the title to the estates passed to John Stackhouse (1741-1819); and specimens from his collection and that of his son, Edward W. W. Pendarves (1778-1853), are in the Russell collection. The wealthy mine adventurer, Robert Hoblyn (?1710-1756) of Nanswhydden, St Columb, was a cousin of Mrs Percival's; long after his death, a fire in 1803 destroyed the house, many historic Stannary records, and his 'whole collection of minerals, amongst which were several unique specimens both of tin and copper.' (Polwhele, 1806, 5:96)

The latter part of the eighteenth century saw an upsurge of interest in mineral collecting. Activity was widespread, throughout Europe and in America, and related to the rapid emergence of mineralogy as a subject of serious scientific study. New chemical elements were being discovered in minerals, and the new science of crystallography found its finest objects of study in the mineral kingdom. The emphasis placed on crystallized specimens is, perhaps, the greatest single difference between the newer collections, from this time onwards, and those of earlier times; and it seems very odd, in retrospect, that the beauty of mineral crystals took so long to be properly appreciated.

From the Cornish standpoint, this sudden growth in mineralogy as a science came at a time of worsening depression in the mining industry. Shortly after 1768, when cheap copper ore began to come from the Parys mine in Anglesey, the Cornish mines – first copper, then tin – went into steep decline, but a little before this, in about 1765, the most celebrated of all Cornish mineral collectors started his activities: Philip Rashleigh (1729-

VERSES *occaſion'd by ſeeing the Foſſilry at* Penderves *in* Cornwall.
Inſcrib'd to Mrs PERCIVAL, *by the Rev Mr* Moore.

ATtend, fair Architect, the Poet's ſong,
To you his ſubject, and his lays belong!
Struck with ſurprize, he views your curious ſtore
Of glitt'ring gems, and variegated ore!
Rang'd with nice art in beautiful array,
The works of *Nature* fairer charms diſplay;
Their dazzling beams refulgent *foſſils* ſhed,
And round the *dome* their various luſtre ſpread.
Di'monds unſpotted, and as chryſtal clear,
In ſtrange variety of forms appear,
Projecting, *ſome* lean forth to meet the ſight,
And *ſome* retir'd, emit a fainter light.
Stones of all colours, and of various ſize,
Diffuſive ſhed their intermingling dies:
Rich veins of glitt'ring *tin*, the rocks unfold,
Or *copper* mark'd with radiant ſtreaks of *gold*;
In diff'rent lights, the diff'rent metals ſhine,
And the *load* runs, as in its *native* mine. (glow,
Above bright *gems*, like ſparkling brilliants,
And light, and beauty, on each other throw;
In faireſt order *all* preſerve their place,
Give and receive a correſponding grace;
Num'rous as ſtars that gild the azure ſkies,
They ſhine, and dazzle each ſpectator's eyes,
When in full ſplendor the bright *god* of day
Darts thro' the radiant *dome* his flaming ray,
Unnumber'd beauties croud upon the view,
And a ſcene opens, wonderful and new;
Then glow the ſparkling gems more clear and
And flaſh around inſufferable light. (bright,
So 'midſt bleak *Zembla*'s plains, with ice o'er-
ſpread,
Some tow'ring mountain rears its gliſt'ring head,
When ſpring returns, the melting *chryſtals* run,
In ſilver ſtreams, and blaze againſt the ſun;
Thro' all the region dart their pointed beams,
With dazzling luſtre, and in fiery gleams.
But *Mundic* over all ſupremely ſhines,
The gayeſt daughter of *Cornubia*'s mines.

In

In diff'rent views ſhe variouſly appears,
And no fixt form, nor certain figure bears.
Promiſcuous colours in her *garb* behold!
She ſhines in *ſilver*, and ſhe flames in *gold*.
Or dreſt in robes of *purple*, *green*, or *blue*,
Her ſev'ral beauties ſhe reveals to view.
Not flow'rs that deck the gaudy *queen of May*,
Nor *Iris' Bow*, more varied dies diſplay.
Mean while delighted, and ſurpriz'd, you hear
New falls of water murm'ring in your ear,
Down a *caſcade*, in gently-flowing tide,
Thro' beds of ore the tinkling waters glide.
How gay! how grand! how beauteous *all* appears,
And the ſtrict teſt of niceſt judgment bears!
Survey this vaſt terraqueous globe around,
A ſcene more brilliant no where can be found!
Let *antient* poets boaſt their *hermits* cells,
Adorn'd with tufted moſs, and vary'd ſhells,
Or *modern* muſes celebrate with praiſe
The ſhell-works and the grots of modern days,
While juſtly you the foremoſt honours claim,
Bright *Manſion*! firſt in beauty, and in fame.
Kedruth, *Dec*. 7, 1755.

Fig. 69 'Verses occasioned by seeing the fossilry at Pendarves in Cornwall' – Rev. Mr. Moore (*Gentlemans Magazine*, 1755).

Fig. 70 Philip Rashleigh's grotto at Menabilly 1806, present whereabouts of painting unknown.

Fig. 71 Weathered specimens lining alcove in garden wall at Clowance House, Praze an Beeble, Cornwall, 1985 (C.V. Smale).

1811), of Menabilly, near Fowey.

Rashleigh's specimens were described by a contemporary as forming

> '[a] rich and magnificent collection, with the inspection of which its truly worthy and liberal possessor has been on all occasions ready to gratify those who study the science.' (Maton, 1797, 1:150).

In retrospect, his collection

> 'was for many years without parallel, both in the County of Cornwall and in fact Great Britain, and as far as many Cornish minerals are concerned will always remain unrivalled.' (Russell, 1952).

Nearly the whole of Rashleigh's collection was bought for the Royal Institution of Cornwall in 1902/3, and is now the pride of the County Museum, Truro. A few more Rashleigh specimens, including some that he figured in his *Specimens of British Minerals* (1797, 1802), remained with other branches of the family until they were acquired by Russell in 1923 and 1958.

In his manuscript catalogue, now at Truro, Rashleigh attached great importance to the accuracy of localities; in another manuscript catalogue, prepared by Aikin in 1814 after Rashleigh's death, some of them are altered without obvious justification. At the front of his catalogue, Rashleigh listed nearly 50 people who had supplied him with specimens, both English and foreign, together with the initials that he used to identify them in the entries (Russell, 1952). The names of Borlase and of William Pryce (?1725-1790), the author of *Mineralogia Cornubiensis* (1778), are notably absent; but of those listed, mention may be made here of Mr John Edwards, Edwd. Fox (of Wadebridge), John Hawkins Esq., Mr William Day, Mr Babington, Rev. William Gregor, Charles Hatchett, Mr Jacob Forster, Mr Maw, Jo[hn] William[s] (jun.) of Scorrier, and Henry Heuland.

Edward Fox (1749-1817), of Wadebridge, was a member of the old and well-known Quaker family who had extensive commercial interests in Cornwall. A merchant, he is not known to have had a mineral collection, but he supplied Rashleigh (through the latter's brother, Charles) with the first-recorded specimens of the 'ore of antimony', later named bournonite, from Wheal Boys (near Wadebridge). Members of the Falmouth branch of the Fox family were concerned with shipping and engineering, and in partnership with the Williams family (see below) and others owned many mining rights. George Croker Fox (d.1807), through his firm G. C. Fox and Sons, started the Perran Foundry in 1791, which grew to great importance in the following century; his brother, Robert Were Fox (1754-1818), was appointed American consul at Falmouth by George Washington. The latter's son, also Robert Were Fox (1789-1877), was a scientist of considerable note; following a patent with Joel Lean in 1812, for the improvement of steam engines, he wrote many papers in the next 50 years on matters relating to mines and minerals. He was a Fellow of the Royal Society, an original member of the Royal Geological Society of Cornwall (in 1814), and founder of the Royal Cornwall Polytechnic Society (in 1833). He, his brother Alfred Fox (1794-1874), and his cousin George Croker Fox (1784-1850) each had small but good mineral collections; these were eventually acquired by (Sir) Arthur Russell, between 1909 and 1929.

Rashleigh's correspondence with other collectors must have been considerable, and we are fortunate that some of the letters that he wrote and received have survived. They shed a fascinating light on the mineral collecting of his time, and the comments they contain on dealers and other collectors show that little has changed since then. There are, for example, complaints about high prices. In a letter (n.d., ?1802) to Rashleigh, John Hawkins lamented that

> 'I have added scarcely any thing worth mentioning to my own collection for to own the truth, my finances now as a married man are not equal to it. The prices in fact of all valuable specimens are become enormous and as I have now so many calls for my money they are fairly out of my reach.'

In similar vein, Rashleigh tells Hawkins (1802 Nov 13) that

> 'The expence of procureing specimens has increased to such extravegance that if my Collection was not considerable at present, I should not have begun now.'

John Hawkins (1761-1841), author, antiquarian, and able scientist, was elected a Fellow of the Royal Society in 1791. In the course of frequent and extensive travels he maintained close contact with scientists in Europe, studying mineralogy under Werner at Freiberg and mineral analysis under Klaproth at Berlin; he supplied the latter with Cornish specimens for investigation, and with information about them (Klaproth, 1787). A founder member of the Royal Geological Society of Cornwall in 1814, he published several papers in its Transactions; and the lists of minerals in the Cornwall and Devon volumes of the Lysons' *Magna Britannia* (1814, 1822) were also his work. His journeys kept him in touch with both Cornwall and London, and a keen interest in horticulture provided another link with a fellow mineral collector, the Rt Hon. Charles Greville (Simmonds, 1942), both men being founder members of the Royal Horticultural Society.

Hawkins' mineral collection was eventually dispersed in 1905, by sale at Foster's auction rooms in London. There were few bidders; some specimens were bought by

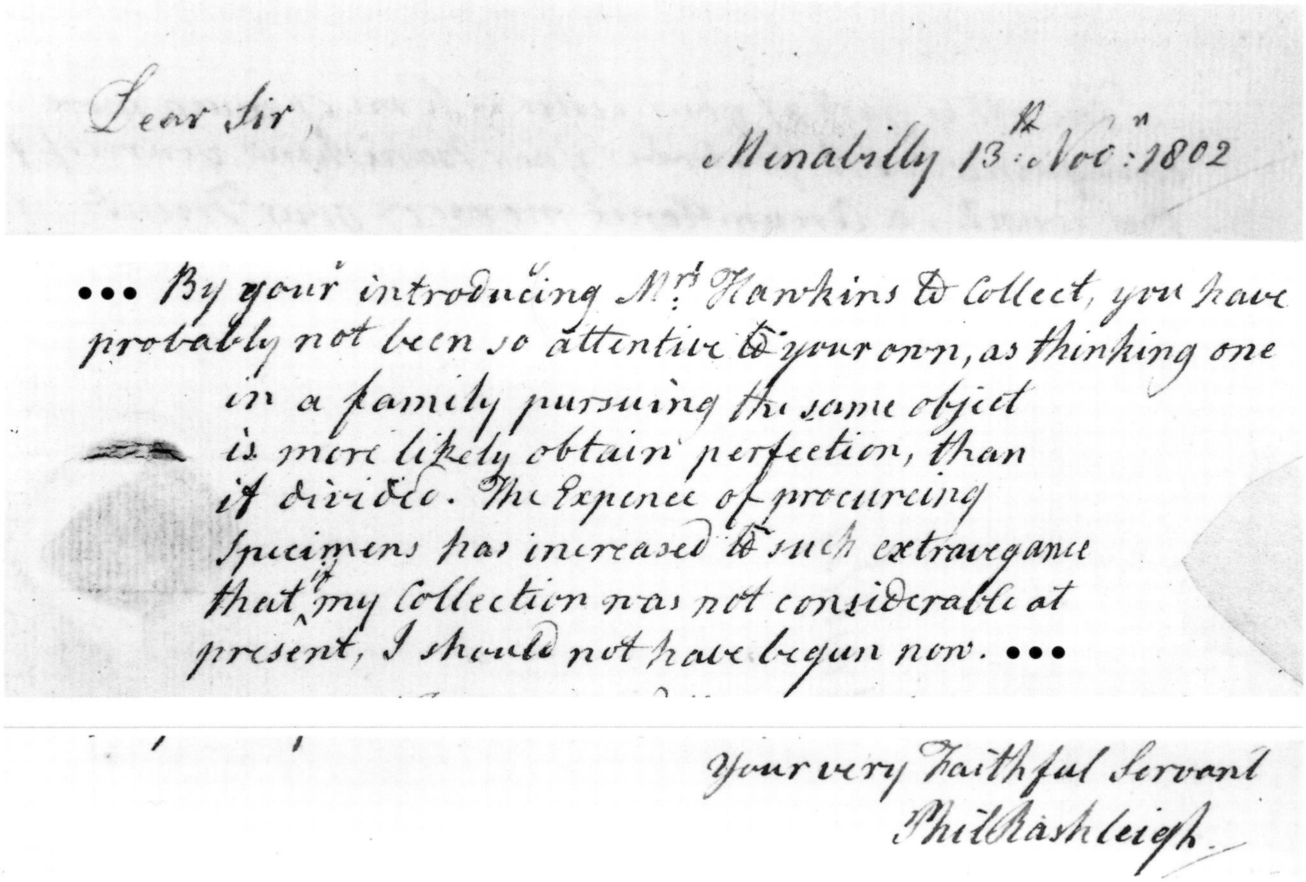

Dear Sir,

Menabilly 13th Nov: 1802

• • • By your introducing Mr Hawkins to Collect, you have probably not been so attentive to your own, as thinking one in a family pursuing the same object is more likely obtain perfection, than if divided. The Expence of procuring Specimens has increased to such extravagance that if my Collection was not considerable at present, I should not have begun now. • • •

Your very Faithful Servant
Phil Rashleigh

Fig. 72 Part of a letter from Philip Rashleigh to John Hawkins, November, 1802.

Russell, but the largest buyer was the German dealer, F. Krantz of Bonn, so that much of the collection left the country (Russell, 1954). Much of Hawkins' correspondence has, fortunately, been preserved and contains some fascinating letters from the discoverer of titanium, the Cornish chemist the Rev. William Gregor (1761-1816; Russell, 1955). Gregor supplied specimens to others, including Rashleigh and Sowerby (see, for example, Tab.CCXLIII, 'Hydrargillite' [= fluellite]), and received American specimens from Archibald Bruce of New York (Paris, 1818:23); but he maintained no proper collection of his own.

Several of the letters to Hawkins (and now in the BM(NH)) are from Joseph Tregoning (1762-1841) of Truro. From around 1791, he was in business as a printer and bookseller, producing (amongst other matter) vols.6-7 of Polwhele's *History of Cornwall.* He also ran a flourishing sideline as a mineral dealer:

> 'we had the opportunity of purchasing excellent Cornish specimens of Mr. Tregoning, an intelligent bookseller of the place, who selects them with judgment, and disposes of them at reasonable prices.' (Warner, 1809:242).

A large advertisement in the *Royal Cornwall Gazette* (31 Jan. 1835) states that he had several thousand specimens in stock, and that he made up small instructional cabinets. From his letters we know that he acted as Hawkins' agent in obtaining specimens, and Russell (*unpub. ms.*) considered him to have been second in esteem only to the celebrated Richard Talling (see below).

Contemporary assessments of some of the principal Cornish mineral collections are contained in the many general histories, accounts of tours, and guidebooks published in the late eighteenth and early nineteenth centuries. Dr J. A. Paris, for example, in his *A guide to the Mount's Bay, and the Land's End* (1816:129), wrote:

> ' . . . before the stranger attempts to purchase any mineral he ought to inspect the several splendid cabinets in the County, these are in the possession of Wm. Rashleigh Esq. M.P. of *Menabilly*, John Williams Esq. *Scorrier House*, Joseph Carne Esq. *Riviere*, . . . [and] that possessed by the Royal Geological Society of Cornwall in their Museum at Penzance. [descriptions follow, to p.134] . . . In order to collect the various minerals of the county the stranger must apply to the different dealers,* and make the best bargain he is able; . . . [The footnote reads:] * The following are the names of the respect-

able dealers to which the mineralogist is recommended, at *Truro*, Tregoning, [and others]'.

In the 2nd ed. (Paris, 1824:130) is added

> 'A miner of the name of *James Wall*, who resides in the village of Carnyorth, has generally a variety of these minerals [axinite etc. from Trewellard and nearby] for sale.').

Joseph Carne (1782-1858) was a banker (in the Batten, Carne and Carne Bank, Penzance), merchant, and for a few years (1807-1819) the resident partner and manager of the Copperhouse (near Hayle) smelter of the Cornish Copper Company. In this latter capacity he succeeded John Edwards (1731-1807), who supplied specimens to Philip Rashleigh. A Fellow of the Royal Society and an honorary member of the Geological Society of London, he was a founder member of the Royal Geological Society of Cornwall and wrote several papers for its Transactions. His fine collection (Paris, 1816:102) went with him to Penzance when the Copperhouse smelter closed (Paris, 1824:31). After his death, a proposal by his daughter Elizabeth (1817-1873) that it should be housed in the museum of the RGSC came to nothing. It was bought in 1899, for the Mineralogical Museum of Cambridge University, but is now, unfortunately, in storage and no longer on display; it contains the original (and for a long time the only) specimen of stokesite, probably obtained from the miner James Wall (Hutchinson, 1900; Couper *et al.*, 1977).

The Royal Geological Society of Cornwall was founded in 1814, in Penzance (Todd, 1964, 1960); from its inception it had displays of the many Cornish and other specimens, listed in its published reports, which were added to its collections by the members. The erection of the present museum, in 1864, marked the start of persistent financial difficulties, which have left both it and the collections in a sad state of disrepair; its future, and the outcome of a restoration appeal (1986), are uncertain.

A later guide to Penzance (Courtney, 1845:38-41) has detailed descriptions of the Royal Geological Society's and the Carne collections, and refers the visitor to the 'cabinets of Messrs Lavin and Edmonds, (dealers in minerals)'. Nothing seems to be known of Edmonds and his dealing activities; perhaps he was Richard Edmonds (1774-1860) or his lawyer son, also Richard (1801-?), a prolific author. We know rather more about John Lavin (d.1856, aged 60), who carried on his mineral dealing in conjunction with the sale of maps, guides, and stationery from premises at nos.6-7, Chapel Street, Penzance, which he grandly styled 'Lavin's Museum'. This remarkable building, in an Egyptian revival style described as 'commercial picturesque' (Curl, 1982:130), still stands to adorn – or, in Sir Arthur Russell's opinion, to disfigure! – Chapel Street. Probably built in about 1836 (Pool, 1974:187), and long neglected, it was faithfully restored in 1973 by the Landmark Trust and is now run as a shop by the National Trust.

Fig. 73 The Egyptian House, Chapel Street, Penzance built by John Lavin for his shop and museum.

Earlier buildings in similar style were Bullock's Museum, Piccadilly, London, and the Civil and Military Library, Devonport, which once housed the St Aubyn collection (see below).

It is not known when Lavin started up as a dealer, but he already called himself a mineralogist when he bought the Chapel Street site in 1835; one of his major suppliers was James Wall, of Carnyorth (above), to whom he owed 'half his fortune' (cited by Barton, 1971:26). He sold instructional cabinets with printed catalogue leaflets as well as individual specimens, and his considerable stock was not confined to Cornish material, but also contained many specimens brought to him by miners visiting home from overseas. Although 'generally and deservedly respected', according to an obituary notice,

> 'It appears that, in common with some other dealers, John Lavin was not above passing off a certain number of faked specimens' (Russell, *unpub. ms.*),

a practice that included mis-localisation; in this way, for example, undoubted Lake Superior coppers came to acquire Cornish labels. A rare example of a faked Wheal

Fig. 74 Artificial specimen, 50 × 22 × 9 mm, faked to resemble a Wheal Coates pseudomorph. The folded sheet-lead has started to swell, due to incipient alteration to white lead, and some of the cassiterite coating has fallen off. The alteration provides an object lesson on the dangers attending the storage of susceptible specimens in drawers made of oak, which releases small amounts of acetic acid into the air.

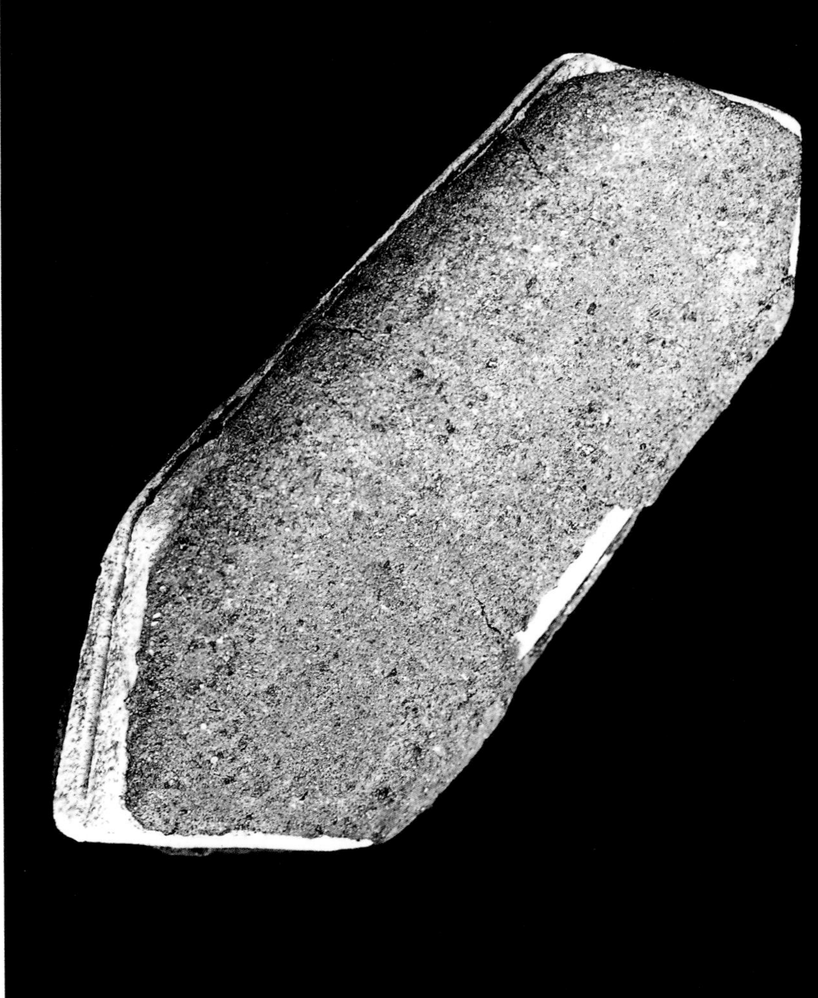

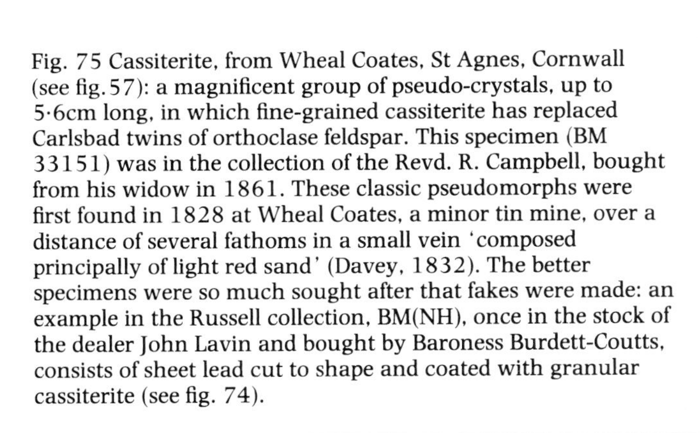

Fig. 75 Cassiterite, from Wheal Coates, St Agnes, Cornwall (see fig. 57): a magnificent group of pseudo-crystals, up to 5·6cm long, in which fine-grained cassiterite has replaced Carlsbad twins of orthoclase feldspar. This specimen (BM 33151) was in the collection of the Revd. R. Campbell, bought from his widow in 1861. These classic pseudomorphs were first found in 1828 at Wheal Coates, a minor tin mine, over a distance of several fathoms in a small vein 'composed principally of light red sand' (Davey, 1832). The better specimens were so much sought after that fakes were made: an example in the Russell collection, BM(NH), once in the stock of the dealer John Lavin and bought by Baroness Burdett-Coutts, consists of sheet lead cut to shape and coated with granular cassiterite (see fig. 74).

Coates pseudomorph, made from sheet lead coated with cassiterite, is among Lavin material in the Russell collection (see photo, below); but, in fairness, it may only have been handled by Lavin and not actually made by him. Later, Greg & Lettsom (1858) expressed their obligation to Mr Lavin (also to Mitchell [? Michell] of Truro, and Garby of Redruth) for information on West Cornwall localities, but by this time John Lavin had died, leaving all his mineralogical effects to his eldest son Edward. Edward soon lost interest in the business, or received an offer that he could not refuse, for in 1863 he sold the entire stock-in-trade to Baroness Burdett-Coutts (1814-1906), the great Victorian philanthropist, for £2500. Her collection was not given to Oxford University, as has been stated, but was bought from her estate in 1922 by Sir Charles Russell, elder brother to (Sir) Arthur Russell. Lavin's celebrated siderite geode, from Wheal Maudlin, was bought at the same time by the Trustees of the British Museum (Natural History), where it is now on display (see fig. 76).

The Williams collection was famous throughout most of the nineteenth century. The prosperity of the mine-owning Williams family, with a confusing number of members named John, began with John [I] Williams, who settled at Burncoose, Gwennap, in 1715 (James, 1949:90-97). His grandson, John [III] (1753-1841), known as 'the King of Gwennap', bought Scorrier House in 1778, which

> 'is known to contain the most valuable variety of Cornish minerals, that was ever collected by any gentleman in Europe.' (Gilbert, 1820:806).

John [III] undoubtedly had some specimens, but the true founder of the collection was his eldest son, John [IV] Williams (1777-1849),

> 'a gentleman certainly one of the most estimable characters, and probably the best chemist, and most experienced mineralogist, in the West of England.' (Warner, 1809:241).

This opinion of his private character, shared by others (James, 1949:92), stands in marked contrast with his later reputation as a ruthless man of business (Barton, 1971:16).

The collection was later moved to Caerhays Castle, bought in 1852 (or 1854) by John [IV]'s brother Michael [II] (1785-1858); some of the drawers had been 'unopened for fifty years' (Maskelyne & von Lang, 1863). It may also have spent some time at Trevince, Gwennap, a house owned by the Beauchamp family but leased to and much improved by Michael [II] (Gilbert, 1820:806).

> 'These unrivalled specimens [cerussite], the pride and glory of the magnificent collection of minerals at Trevince, were brought from this place [Pentireglaze mine]. Some of the crystals, finer than the hair of the head and silvery white, were as much as nine inches long. Two men were hired to carry single specimens in a basket slung on a pole resting on their shoulders to prevent their being injured, a distance of upwards 30 miles – well worth the trouble and expense, so valuable were they.' (*Mining J.*, 30 Jan. 1858; quoted by Jenkin, 1970(XVI):22).

John Michael Williams (1813-1880), second son of Michael [II], was born at Trevince and lived at Pengreep, another Beauchamp property, before inheriting Caerhays. He added to the collection, in which he was assisted (in the 1860s) by one of his clerks at the Burncoose office, William Semmons (1841-1915). Semmons transferred to the Williams office in Liverpool, becoming manager, and finally became a self-employed metal broker in London. As a mineral dealer, he sold many fine Cornish specimens to European dealers, including Krantz of Bonn and Bohn of Vienna; on his death, much of his excellent collection was bought by (Sir) Arthur Russell (*unpub. ms.*).

Semmons also learned much from John Garby (1812-1864) of Redruth, one of Greg and Lettsom's sources of information. Garby, who was mine purser at Wheal Basset at one time, died in Brazil on a prospecting trip. Of his several publications, perhaps the most useful is a comprehensive list of Cornish minerals (Garby, 1848). His fine collection, neatly labelled, became dispersed: the first selection went to his friend W. W. Smyth, and was later acquired by Sir Arthur Russell (see metatorbernite photo, below); other specimens went to the Redruth School of Mines, and most of the remainder to Richard Talling (see fig. 77), the dealer. Another dealer, whose customers included John Michael Williams and Richard Talling, was Thomas Light Richards (1818-1887) of Lanner Moor, Gwennap; his son, G. H. Richards, was also a dealer, in London (Russell, *unpub. ms.*).

Family interest in the Williams collection, which – surprisingly – had no catalogue and was largely unlabelled, appears to have ended at John Michael's death; and it was eventually dispersed in somewhat piecemeal fashion. In 1893, John Charles Williams (1861-1939), John Michael's son, invited L. Fletcher and H. A. Miers to Caerhays to select 550 specimens for the British Museum (Natural History), amongst them a fine spangolite (Miers, 1894); several thousand specimens had already been given to the Basset Memorial Museum of the Camborne School of Mines, in 1891 (Anon, 1892:213; Piper, 1974:26). The pick of another part of the collections, which had been kept aside in the Estate Office at Burncoose, was acquired by Sir Arthur Russell in 1949.

The Camborne School of Mines was a surprisingly young institution when it received the Williams minerals, having been founded only 3 years earlier in 1888

(Piper, 1974:16). The technical education of miners, however, had already been under discussion for most of the century; instructional classes organized by the Mining (or Miners') Association, in conjunction with the Schools of Science and Arts in Camborne, Redruth, Penzance, and elsewhere in the county were well attended. The Mining Association and Institute was absorbed by the Camborne School of Mines in 1895, and smaller Schools of Mines, at Penzance and Redruth, were closed and transferred with their mineral collections to Camborne in 1909. The Camborne School and collections were moved from the centre of Camborne to their present site, beside the main (A30) road to Redruth, in the mid-1970s.

The collections of the Penzance school were undoubtedly the work of the Principal, Andrew Ketchan Barnett (1852-1914), who was mayor of the town for some years and was also a dealer in minerals (Spencer, 1921:241). Other Redruth collections were in the Robert Hunt Memorial Museum, founded in 1889; Robert Hunt (1807-1887) was a pioneer investigator of photographic processes, a keen mining educator who created the Miners' Association of Cornwall and Devonshire, and Keeper of the Mining Records Office (Pearson, 1976). An early lecturer of the Miners' Association was (Sir) Clement le Neve Foster (1841-1904); a member of the Geological Survey 1860-65, he was Inspector of Mines for Cornwall and Devon 1872-80. His mineral collection, with good Cornish representation, went to the Royal School of Mines, London, where he was Professor of Mining from 1890.

Seventeenth- and eighteenth-century London abounded in all manner of private collections and museums (Altick, 1978), many containing rock and mineral curiosities. By the latter half of the eighteenth century, serious private mineral collecting in the capital was well established: William Hunter (1717-1783), the famous surgeon, had Cornish material among the several thousand minerals in his collections at Great Windmill Street, which he left to Glasgow University (G. Durant, *priv. comm.*). Institutional collections at this time, on the other hand, were dormant; and those of the Royal Society were given to the British Museum in 1781.

In 1799, a committee consisting of Philip Rashleigh (on one of his infrequent trips outside Cornwall), the Rt Hon. C. F. Greville, and Sir Joseph Banks assessed the British Museum's minerals and commented on the representation of British localities as being poor, and much worse than visitors had a right to expect. As a first step towards improvement, they recommended the purchase of Hatchett's collection of 7000 specimens, for £700. This contained material from Cornwall and Devon, much of it probably collected when he accompanied W. G. Maton on his tours in 1794 and 1796 (Maton, 1797; Raistrick 1967); some may have come from Rashleigh, to whose collection he had contributed specimens. Charles Hatchett (1765-1847), an able chemist of independent means, was the discoverer of the element columbium (niobium) in one of Sloane's specimens at the British Museum; he encouraged his protégé W. T. Brande (see below), later his son-in-law, to take an interest in mineralogy (Weeks & Leicester, 1968:326). Little can now be found of Hatchett's collection, nor of Greville's which was bought in 1810.

The Rt Hon. Charles F. Greville (1749-1809), whose home at Paddington Green has long disappeared, had an excellent mineral collection (Campbell Smith, 1969; Simmonds, 1942). The Count de Bournon (1751-1825), who arrived in England in 1794 as a refugee from the terrors of the French Revolution, having lost all of his possessions (including his mineral collection), was helped financially by Greville and Sir John St Aubyn (1758-1839), and to a less extent by Sir Abraham Hume (1748-1838), in return for looking after and cataloguing the minerals in their collections (Bournon, 1813; Dellow, 1973:18). Bournon's catalogue of the Greville collection has not survived.

Sir John St Aubyn, a partner in the Cornish Copper Company and whose family estates were based at Clowance, in Cornwall (above), spent much of his time in London, at Lime Grove in Putney. We know neither the size nor detailed content of his collection which was, apparently, split into two in 1834 (Torrens, 1977). The larger part, housed at various locations in Devonport until 1924, is now in the Plymouth City Museum (Curry, 1975), together with the first two volumes of Bournon's catalogue; possibly all that were completed, these stop well short of the 'metallic' and secondary Cornish minerals. Between 1795 and 1799, St Aubyn had purchased for £3000 the collection of Dr William Babington (1756-1833), most of the specimens in which were from the 'Butean Collection' of John Stuart (1719-1792), Third Earl of Bute; but, considering Babington's contact with Rashleigh, his published catalogue (1799) contains surprisingly few entries for Cornwall or Devon.

After selling his collection, Babington maintained his interest in the subject: he was a member of the short-lived (1799-1806) British Mineralogical Society (Weindling, 1983), and was a founder member of the Geological Society of London in 1807 (Woodward, 1907). Bournon's own collection returned with him to France, in 1815, and is now at the Collège de France; he gave various specimens, some Cornish, to the Abbé Haüy and these are now in the Paris Natural History Museum. Bournon had early connections with America, through the collector Archibald Bruce (Greene & Burke, 1978).

Greville, St Aubyn, and Hume, doubtless with stage management by Bournon, met in 1804 to discuss the formation of a 'National Collection of Mineralogy', because '. . . Mineralogy requires a distinct Establishment ...'. Following further discussions with others, including Charles Hatchett, the Royal Institution (of London)

Fig. 76 Siderite, from Wheal Maudlin, Lanlivery, Cornwall (see fig 61): short prismatic crystals with basal plane, the majority about 13 mm across and all showing concentric zoning in shades of brown, with milky terminated prisms of quartz, on pyrite lining part of a large geode (33 × 20 cm and 12 cm deep).
Little more than a mile from the much larger Restormel iron mine, Wheal Maudlin had a small reported output of copper and tin; it is best known among mineralogists for these highly-characteristic, siderite crystals, found in about 1820. It is quite possible that the many known smaller specimens were all from other parts of this single cavity.
This remarkable specimen (BM 1922,314) was bought in 1922 from the executors of Baroness Burdett-Coutts, who had acquired it in 1863 with the rest of the collection and stock-in-trade of the Penzance dealer, John Lavin. It is mentioned by Greg and Lettsom (1858:259), when it was still in Lavin's possession, but there is an earlier reference in a letter to John Hawkins from the Truro dealer, Joseph Tregoning (11 Dec. 1834; BM(NH) Library): '. . . a peculiar variety of Carbonate of Iron, found about 14 or 15 years ago in a mine called Maudlin, between Bodmin and Lostwithiel, and long since ceased to work. The specimens are now extremely scarce and much sought after. I have not seen for 12 years a single specimen . . . One specimen in Penzance for sale is valued at £150!'

adopted the idea and started its own collections (Berman, 1978:89). Sir Humphry Davy (1778-1829), the famous Cornish chemist and professor at the Royal Institution, was active in adding specimens, and its growth was continued by W. T. Brande (1788-1866) who published a catalogue in 1816. In spite of their having raised some of the money required for the project, Greville and his two friends were denied access to the Royal Institution collections (Berman, 1978:91), so it is scarcely surprising that their own collections went elsewhere.

> 'The [Royal Institution] collection was subsequently broken up and distributed amongst various institutions. A large part went to the Jermyn Street Museum.' (Murray, 1904, 2:351).

The 'Jermyn Street Museum', or Museum of Practical Geology, was removed to South Kensington in the 1930s and became the Geological Museum; it came under the administrative control of the British Museum (Natural History) in April 1985. Murray must have seen documents relating to the transfer of the Royal Institution's collection, but they cannot now be found at either establishment; nor can individual specimens from it be identified.

Mr Jacob Forster, Mr Maw, and Henry Heuland, listed by Rashleigh, were major dealers with an international clientèle. (Adolarius) Jacob Forster (1739-1806), formerly professor of mineralogy at St Petersburg (where he also spent the last years of his life), had business premises there, in Paris (run by his brother Henry), and in London (run by his wife Elizabeth, in Gerard Street, Soho) (Frondel, 1972; Whitehead, 1973). Elizabeth's brother, George Humphrey (?1739-1826), also dealt in minerals and other natural history specimens, mainly shells, from premises in Leicester Square. John Henry Heuland (1778-1856) (Russell, 1950) was Jacob Forster's nephew; the Forster–Heuland–Humphrey family relationships are complex (Vallance, 1981). Forster's private collection, with additions, was sold by Heuland to C. H. Turner in 1820. Armand Lévy, then living in London, was commissioned by Heuland to write a descriptive catalogue; as a source of original information on minerals from Cornwall and Devon, this would have been more important had its preparation for publication not been delayed until 1838. Bought by H. Ludlam (see below), these specimens are now in the Geological Museum, London.

Fig. 77 Metatorbernite from the Old Gunnislake mine, Calstock, Cornwall (see fig. 64): a composite rosette, 5cm across, of square platy crystals, on a matrix of altered granite. The crystals are clear, lacking the usual cloudiness of metatorbernite resulting from the natural dehydration of torbernite, and this suggests that they probably crystallized at this locality in their present 'meta' (lower hydrate) state. Originally considered to be a copper ore, and later an oxide of uranium, the first Cornish specimens of torbernite were found near St Day and Redruth; the discovery at Old Gunnislake, a copper mine at the extreme eastern edge of Cornwall, was in 1811/12 and at a depth of 90 fathoms from the surface. Identification of the different lodes is now impossible, but this mine or a neighbour also produced excellent specimens of chalcophyllite ('tamarite') and other copper species. Until they were moved in the early 1970s, becoming the base of a football pitch, the sandy dumps in Gunnislake village could always be relied on – especially after a spell of rainy weather – to yield small specimens of metatorbernite.
This specimen (BM 1964R,7465) is in the Russell collection; it was once in that of Warington W. Smyth, who was given it by J. Garby.

Heuland's own private collection was dispersed; its catalogue, at one time in Calvert's possession, has disappeared, but a selective transcript is at the BM(NH). John Frederick Calvert (d.1897), amongst other things a mining engineer, may (or may not) have accompanied Heuland on collecting trips in Europe. Calvert travelled extensively, obsessed with the search for gold, and has been credited with finding the Belstone Consols copper mine, Devon (Barton, 1971:43). Said to have been known as 'Lying Jack', almost nothing written by or about Calvert is free from doubt or exaggeration, starting with his year of birth (1811, or 1814, or even 1825 (T. G. Vallance, *priv. comm.*)) and the claim that he was acquainted with Philip Rashleigh (who died in 1811). Few specimens survive to support the claim that his mineral collection was superlative; it was sold off piecemeal, and the residue was eventually bought by the American dealer, Martin Ehrmann, in 1938.

On his tour in 1802, E. T. Svedenstierna recommended four London dealers (in addition to Forster in Soho)

> '[who] also deal in birds, snails, cut stones – in short, with all kinds of natural curiosities and oddities'

as the best mineral dealers in London (Dellow, 1973:20). George Humphrey is not among them; nor is Rashleigh's 'Mr Maw', who only started his London shop, at 149 Strand, in 1811, the year of Rashleigh's death. John Mawe (1764-1829) was a native of Derbyshire, where he must already have been a dealer at the close of the century when he was commissioned to collect British minerals for the King of Spain. With a royal appointment as mineralogist to the King of Portugal, he spent the years 1804-10 in Brazil. His most interesting contribution to the mineralogy of south-west England is a splendidly circumstantial account of the classic tourmaline (and apatite) find near Bovey Tracey, Devon (Mawe, 1818). After Mawe's death, his business was carried on by James Tennant (?1808-1881), who also lectured in mineralogy at King's College nearby.

Svedenstierna had introductions from Bournon, and after describing Greville's collection in some detail he went on to note that St Aubyn's

> 'is not so complete as that of Mr Greville, but yet contains a quantity of fine specimens, particularly from Cornwall . . . Among the smaller, but rather good mineral collections are: (1) that of Mr Richard Phillips, Georges Church Yard, which deserves to be seen especially on account of its fine copper specimens, rich tin ores, etc., from Cornwall . . . (2) Mr Sowerby's collection, Mead Place, Lambeth, contains a great many crystals, and is daily enriched with new kinds. Mr Sowerby is an engraver, and has actually founded this collection for the purpose of a work on the minerals of England which he wants to publish ...' (Dellow, 1973:19).

Richard Phillips (1778-1851), an accomplished chemist and mineralogist, was from 1839 until his death the first curator of the Museum of Practical Geology, in Jermyn Street; and, for lack of any information to the contrary, we may presume that his collection went there. His address, George Yard, off Lombard Street, was that of the family printing business run by his elder brother, William Phillips (1775-1828). Their father, James Phillips, had printed Pryce's *Mineralogia Cornubiensis* in 1778, and we may guess that this had some influence on their interest in minerals. William Phillips was a noted mineralogist and crystallographer, considered (Woodward, 1907) to be the most distinguished, as a geologist, of the founders of the Geological Society of London. His collection of crystals, sold by G. B. Sowerby by private contract, was given by Dr Rutter to the Medical Institution of Liverpool; transferred to the Liverpool Museum in 1877, it was destroyed in an air raid in 1941.

William Phillips' paper on cuprite (1811) contains the best contemporary account of the veins at Wheal Gorland. The illustrations in this, and in his exhaustive account of the crystallography of cassiterite (1814), provide examples of the work of Wilson Lowry (1762-1830), the artist and draughtsman who pioneered steel-plate engraving. Lowry's mineral collection, highly regarded by Bournon (1813:150), was dispersed at auction in 1830.

James Sowerby (1757-1822) excelled as a natural history artist and publisher, with interests in botany, mineralogy, and conchology. His principal works on minerals, as with his other publications, were issued to subscribers in separate parts for subsequent assembly: *British Mineralogy* appeared in five volumes, between 1804 and 1817, and most of the 550 plates are dated. Sowerby had extensive collections of shells and fossils, and his son G. B. Sowerby was a dealer, but the text indicates that most of the specimens that he figured in *British Mineralogy* belonged to other collectors. One of these was William Day, a contributor to Rashleigh's collection and whose own fine collection may have been acquired by Sowerby, but of whom nothing seems to be known. The extent and fate of Sowerby's mineral collection, as distinct from his shells and fossils, is also unknown.

Another of the collectors, with specimens figured by Sowerby in 1816 (vol.5, Tabs.DXVIII, DXXIX), was ' - Lounds, Esq.', (or Lowndes). William Lowndes (1752-1828), a Chief Commissioner of Taxes, had a splendid mineral collection which was sold at auction in 1828. His 'nail-head' chalcosine, from Wheal Abraham, was bought by Joseph Neeld (1789-1856), who in 1827 had inherited the fortune of his uncle Philip Rundell, partner in the London gold- and silversmiths Rundell and Bridge; Neeld's minerals were bought in 1974 (B. Lloyd, *priv. comm.*), and the chalcosine (pictured below) is now in the BM(NH).

One of the best-known London dealers in the mid-nineteenth century was Bryce McMurdo Wright (d.1874, aged 60), with premises close to the British Museum at 90, Great Russell Street. A native of Caldbeck, he advised Greg and Lettsom on Cumberland localities; but, although he is known to have sold many fine Cornish specimens, he appears to have been a secondary source, buying from other dealers. Wright's son, with the same names as his father, continued to run the firm for a few more years.

The London firm of mineral dealers with by far the longest continuous history was founded by James Reynolds Gregory (1832-1899). Started in about 1850 as J. R. Gregory and Co., and continued by his son Albert G. F. Gregory, the name changed to Gregory, Bottley and Co. in 1932 when E. Percy Bottley (1904-1980) became a partner. In 1981 the firm was bought by Brian Lloyd, for the previous seven years a dealer in his own right, and continues to trade as Gregory, Bottley & Lloyd. Over this period the firm absorbed several others, including those of Russell and Shaw (founded in 1848, by Thomas Russell); Samuel Henson (1848-1930); F. H. Butler (1849-1935); and G. H. Richards (above).

Henson's father, Robert Henson (1814-1864), had worked for James Tennant (above) before setting up his own business in 1840 (Spencer, 1930:395). Francis Henry Butler, at one time an assistant editor of the *Encyclopaedia Britannica* (9th ed.), was a qualified medical practitioner, acting as a 'locum' for neighbouring doctors as well as being a mineral dealer (Spencer, 1936:279). Butler was executor to Richard Talling, whose remaining specimens formed the basis of the teaching collections in which he specialized, at his shop 'The Natural History Agency', near the Natural History Museum in South Kensington.

Richard Talling (1820-1883) was, beyond question, the greatest Cornish dealer of all time, flourishing in the years when mining activity was at its peak in Cornwall

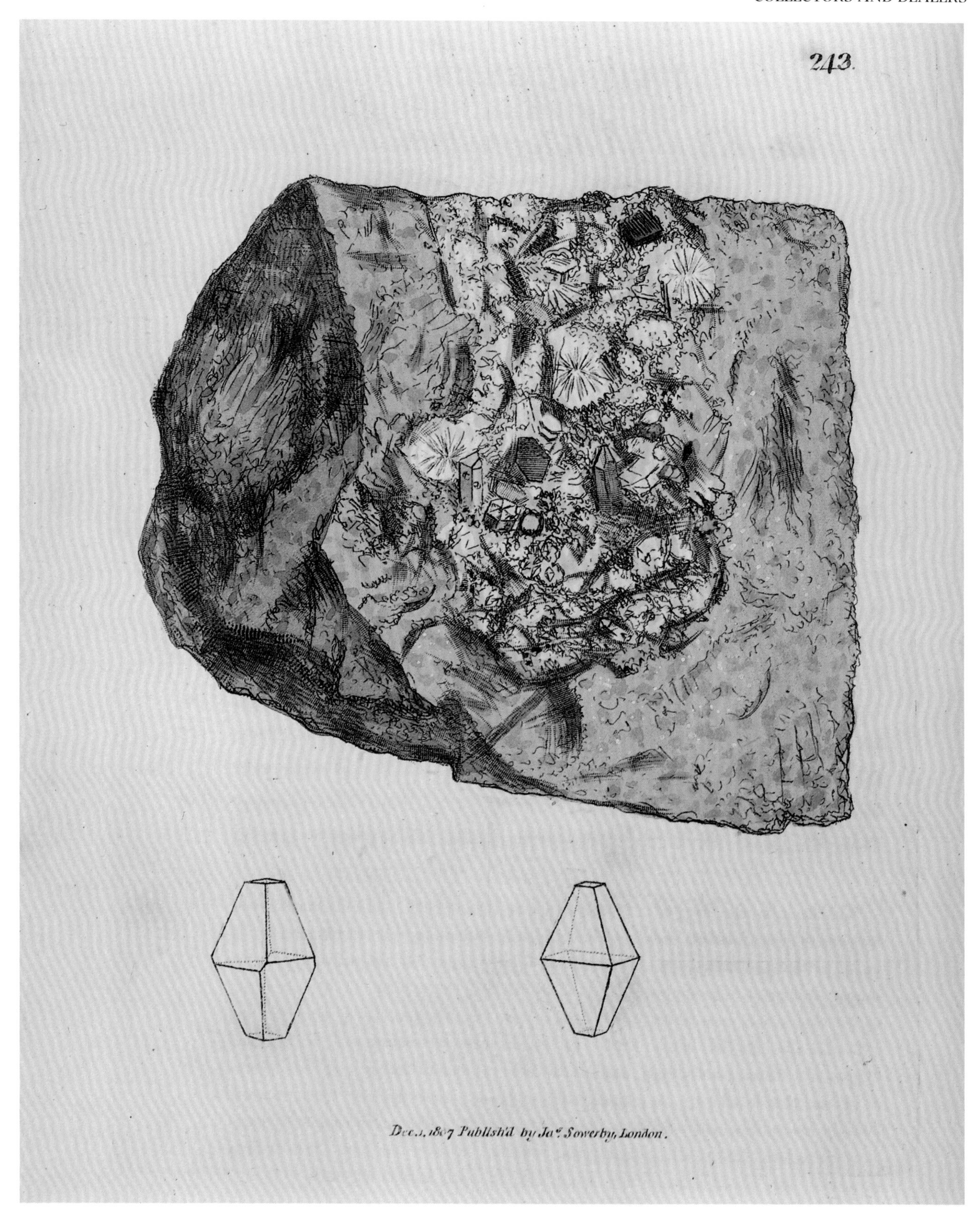

Fig. 78 A James Sowerby coloured figure (No. 243) of hydrargillite (later shown to be fluellite) from the Stennagwyn mine, St Stephen-in-Brannel, Cornwall.

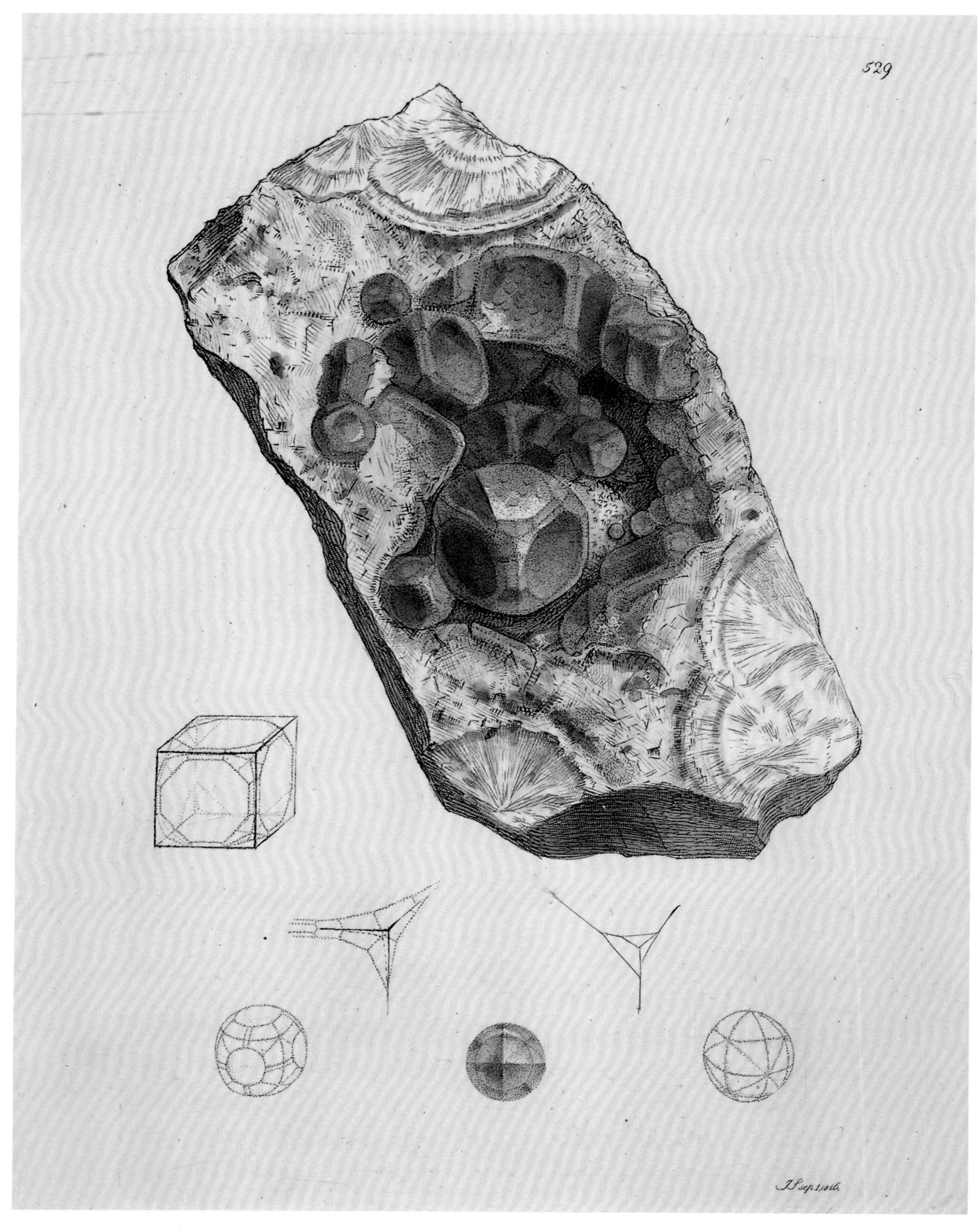

Fig. 79 A James Sowerby coloured figure (1817. No. 529) of zoned fluorite with rounded corners from the Bere-Alston lead mine in Devonshire (Lent to Sowerby by W. Lowndes.)

and Devon. He started business in or before 1844, and in 20 years had become the

> 'dealer from whom the National Collection [British Museum] has received a very large proportion of its finest Cornish minerals' (Maskelyne, 1865).

His shop, at no.18, Fore Street, Lostwithiel, which he called a 'Fancy Repository' (selling stationery, ornaments, trinkets, and gift items), was the base from which he scoured the two counties and handled many of the best specimens they produced. Among these were fluorite from West Caradon (1845-1867); childrenite from George and Charlotte, Devon (1846 +); fluellite from Stennagwyn (1847); fluorite and baryte from Menheniot (1850-65); goethite from Restormel (1854-56); bournonite and tetrahedrite from Herodsfoot (1858-68); calcite from Wheal Wrey (1861-65); cuprite from Wheal Phoenix (1858, 1864-74); chalcotrichite and langite from Fowey Consols (c.1864); vivianite from Wheal Jane (1867 +); liskeardite from Marke Valley (1874); ludlamite from Wheal Jane (1876); and many others.

Although thoroughly commercial in his dealings, which led him (in common with many other dealers over the years) to be very secretive or misleading over localities, Talling had a keen eye for the unusual. An excellent example of this is the mineral he took for eulytine, which has recently proved to be crystallized waylandite (Clark *et al.*, 1986); and his danalite from an undisclosed locality in Cornwall, found in about 1864, would have been a new species earlier had it been analysed at the time

Fig. 80 Chalcosine, from Cornwall: grey-black pseudo-hexagonal crystals, the largest 11 mm across, perched on columnar aggregates of the same. This habit variety of chalcosine, at one time known as 'nailhead copper ore' from its appearance, is very uncommon. This specimen (BM 1976, 246) was bought in 1976 from the dealer Mr Brian Lloyd, who acquired it as part of the Neeld collection formed in the early nineteenth century. It is so closely similar to the one owned by '-Lounds Esq.', figured by James Sowerby in his *British Mineralogy* (1817, vol. 5, Tab. DXVIII), that it may well be the same; so it seems safe to assume that the correct locality is Wheal Abraham (see fig. 40), and that it was found in 1813. 'Lounds' was William Lowndes (1752–1828), whose mineral collection was sold at auction in 1828. Wheal Abraham, later joined with its neighbour and worked as Crenver and Wheal Abraham, was a prosperous copper mine in Crowan parish, about 4 miles from Camborne.

(Miers & Prior, 1892; Kingsbury, 1961). He was directly responsible for the discovery, leading to their description as new species, of churchite, langite, botallackite, tavistockite (now discredited), andrewsite, and ludlamite; tallingite was named for him, but has been shown to be connellite (Bannister *et al.*, 1950).

Talling secured for posterity most of the magnificent bournonite specimens from the Herodsfoot mine, and even if this had been his sole achievement it would have earned him an honoured place in mineralogical history. His persistent trafficking caused the mine manager, Captain T. Trevillion, to impose a ban, which Talling circumvented by buying a share in the mine (Russell, *unpub. ms.*). He is not known to have had a private collection, nor to have left any notes: but invoices and correspondence from him are preserved at the BM(NH); in the Ludlam papers at the Geological Museum; and in the J. A. Clay (of Philadelphia) papers at the Smithsonian Institution (P. E. Desautels, *priv. comm.*).

Talling's principal customer, next to the British Museum, was Henry Ludlam (1822-1880), a wealthy hosier who lived in Jermyn Street, close to the Museum of Practical Geology. He also bought the Turner (ex-Heuland) and Neville collections. Sensitive to slights about his amateur status, Ludlam's closest scientific friends were the curators Thomas Davies of the British Museum, and F. W. Rudler of the Museum of Practical Geology. His specimens, over 18000 in all, were largely unlabelled and uncatalogued at the time of his unexpected death, and the task of documentation fell to Rudler. Selected material from this splendid collection, bequeathed to the Museum of Practical Geology, remains on display in the Geological Museum at South Kensington (Rudler 1905).

Excluding only the merger with the Geological Museum, in 1985, the largest single addition ever made to the minerals of the British Museum (Natural History) was the Russell collection of British minerals, acquired by bequest in 1964. Arthur Edward Ian Montagu Russell (1878-1964), encouraged by his mother, began to collect minerals while still a child; he maintained a passionate interest in them throughout his life, and in addition to maintaining familiarity with dumps and surface localities he made underground visits to almost every metalliferous mine in the British Isles. His earlier years of employment were spent with the old London and South Western Railway, but later he was occupied with the running of the family estate at Swallowfield Park, near Reading; he succeeded to the family title, becoming Sir Arthur, on the death of his elder brother in 1948. As well as collecting in the field, Russell was assiduous in tracking down and purchasing collections, both old and contemporary, selecting specimens that he needed and selling the remainder to finance his hobby; collectors and dealers among his American trading contacts included Arthur Montgomery and Hugh Ford. Some of the old collections that he bought have been mentioned above, and in an obituary (Kingsbury, 1966:675), but the list of even the southwest collections is too long for this account. Reference must be made, however, to one of his early (1911) acquisitions, the Pearse collection, both for its intrinsic interest and because Devon has received little mention. Edmund Pearse (1788-1856), a surgeon of Tavistock, made a collection of fine specimens from the Virtuous Lady mine which, together with his manuscript catalogue, are nearly all in the Russell collection (Russell, 1913).

One of the kindest and friendliest of men, Sir Arthur Russell was always happy to welcome visitors wishing to see his collection in the mineral room at Swallowfield, and he went out of his way to encourage their interest. One of his many close friends, who owed much to this encouragement, was Arthur William Gerald Kingsbury (1906-1968), whose collection is now also in the British Museum (Natural History).

There remain good opportunities for collecting in Cornwall and Devon at the present day, despite the removal of so many productive dumps and the survival of so few mines. Interesting discoveries are being made, not only on old specimens, and there are several dealers and private collectors still active in the area. Richard William Barstow (1947-1982) was, perhaps, the most enterprising of these; and, in 1986, the Plymouth City Museum bought from his widow a selection of the fine specimens from Cornwall and Devon that he was able to obtain for his private collection during his regrettably short life.

Tetrahedrite, from the Herodsfoot mine, Liskeard, Cornwall: a group of modified tetrahedra (individual crystals up to 1·8cm) encrusted with chalcopyrite on quartz and galena. This specimen (BM 32864) was bought from Mr Richard Talling in 1861.

THE MINERALS

Acanthite, from Wheal Newton, Harrowbarrow, near Calstock, Cornwall (see fig. 64): unusually large octahedra, with edges up to 18mm, in parallel position. These are pseudo-crystals, paramorphous after argentite which is only stable at higher temperatures.
Wheal Newton, earlier named the Harrowbarrow mine, was worked for lead, silver, and copper in the third quarter of the 1800s. The intersections of the east-to-west copper lodes and the north-to-south lead lodes were particularly rich in silver ores, but few specimens survived. This specimen (BM 62852) was bought from the dealer Samuel Henson in 1887.

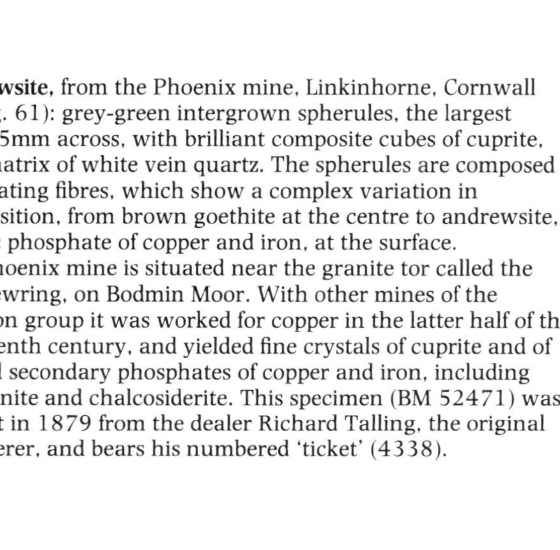

Andrewsite, from the Phoenix mine, Linkinhorne, Cornwall (see fig. 61): grey-green intergrown spherules, the largest about 5mm across, with brilliant composite cubes of cuprite, on a matrix of white vein quartz. The spherules are composed of radiating fibres, which show a complex variation in composition, from brown goethite at the centre to andrewsite, a basic phosphate of copper and iron, at the surface.
The Phoenix mine is situated near the granite tor called the Cheesewring, on Bodmin Moor. With other mines of the Caradon group it was worked for copper in the latter half of the nineteenth century, and yielded fine crystals of cuprite and of several secondary phosphates of copper and iron, including libethenite and chalcosiderite. This specimen (BM 52471) was bought in 1879 from the dealer Richard Talling, the original discoverer, and bears his numbered 'ticket' (4338).

Apatite, from Luxulyan, Cornwall (see fig. 61): doubly-terminated blue-green prisms, the largest 37 × 15mm, with gilbertite hydromica on quartz-feldspar pegmatite. The Colcerrow granite quarry at Luxulyan, where it was being worked in the nineteenth century, produced some of the finest apatite crystals known from Cornwall. This specimen (BM 67264) is probably from Colcerrow, and was bought from the collector William Semmons in 1891.

Apatite, from Woolley farm, near Bovey Tracey, Devon (see fig. 64): a large (28 × 40mm) sharply terminated prism with smaller individuals, accompanied by black tourmaline (schorlite) and brown altered feldspar, on granite. Schorlite crystals from this locality are well represented in old collections, but the apatite is less well known because most of the crystals were destroyed by ignorant labourers. Exceptionally, the apatite prisms were just over 5cm long. As far as can be ascertained, they all came from a single pegmatitic cavity of the Dartmoor granite, which was uncovered during ploughing, a year or two before 1817. Both minerals were figured by Sowerby (*British Mineralogy*, 1817, **5**: Tabs. DXLVIII, DXLIX), and a fascinatingly circumstantial account of the discovery was written by the dealer John Mawe in 1818. This specimen (BM 72058), formerly in the collection of the Williams family, Cornish mine owners, was donated by Mr J. C. Williams in 1893.

Arsenopyrite, from Wheal Penrose, Sithney, Cornwall (see fig. 40): tiny (0.25mm dia.) stellate repeated twins (trillings), with yellowish dolomite crystals and dark-brown sphalerite, in cavities of vein quartz.
This specimen (BM 1971,307) is one of several found on the old mine dumps by a private collector, Mr Richard Sparks, and was donated by him in 1971.

Arsenopyrite, from the Virtuous Lady mine, Buckland Monachorum, Devon (see fig. 64): tapering intergrown crystals, up to 3·2cm long, with a bronzed tarnish and bounded by faces of {101} and {120}; brown globular aggregates of siderite are attached to a few of the arsenopyrite edges, and there is a little encrusting green scorodite.
This magnificent specimen (BM 20204) was bought from Mr Richard Talling in June, 1846, early in his career as a dealer; one of the first specimens that he sold to the BM, at the end of 1844, was a large group of lenticular siderite crystals, also from the Virtuous Lady mine.

Axinite, from the Stamps and Jowl Zawn, Roscommon Cliff, St Just, Cornwall (see fig. 40): sharp-edged composite crystals up to 26mm across, of the textbook clove-brown colour, on finer-grained matrix. Fine crystals like this, which merit comparison with those from the classic locality in Dauphiné, were first found at this locality a year or two before 1821; inferior crystals were known from St Just much earlier. This recent specimen (BM 1969,286) is in the Kingsbury collection, at the BM(NH), and was collected *in situ* by Mr Arthur Kingsbury in 1952.
Several veins containing crystallized axinite occur here in a calc-silicate hornfels, formed by alteration of volcanic rocks adjacent to the Land's End granite mass. The term 'zawn' is confined to the Land's End area; it denotes a steep-sided inlet, flanked by hard rock, which has been caused by the action of the sea in preferentially eroding the softer rock of a mineralized vein or fissure.

Baryte, from Wheal Mary Ann, Menheniot, Cornwall (see fig. 61): pale tan bevelled tabular crystals, up to 5cm cross, with clear orange-brown patches, on a twinned fluorite cube and associated with a little quartz and galena.
Wheal Mary Ann, now long abandoned, started work in 1846 and was one of several prosperous lead mines in the neighbourhood of Liskeard. In marked contrast to its abundance in the North of England, baryte is not very common in Devon and is quite rare as a gangue mineral in Cornwall; specimens from the other notable Cornish locality, the ancient Ale and Cakes mine, Gwennap, are in many old collections. Wheal Mary Ann also produced exceptional fluorite specimens, in zoned cubes up to 25cm edge, often showing the {421} form to perfection.
This specimen (BM 34263) was bought from Mr Richard Talling in 1862, and there are other fine examples in the Geological Museum, London.

Bassetite, from the Basset mines, Illogan, Cornwall (see fig. 46): small yellowish-brown crystal plates, the longest edge about 2mm, in parallel growths on cavernous 'gossan'.
Long mistaken for the related species autunite, bassetite was characterized and named for the locality in 1915. This specimen (BM 1963,478) was presented by Arthur Kingsbury.

Bournonite, from the Herodsfoot mine, Liskeard, Cornwall (see fig. 61): a group of bright grey, simply-twinned crystals surrounded by quartz. The largest face is about 1cm long.
This specimen (BM 34307) was bought from Mr Richard Talling in 1862, and is remarkable in revealing the true symmetry of the mineral. Inspection of the faces enclosing the re-entrant angle of the twins shows that they are at right angles to adjacent faces, and the edges are not modified, thus indicating that there is no mirror plane of symmetry parallel to them. This feature was overlooked in earlier studies.

Bismuthinite, from the Fowey Consols mines, St Blazey, Cornwall (see fig. 61): slender iridescent prisms, up to 25mm long, on and penetrating both chalcopyrite crystals and vein quartz.
Free-standing crystals of bismuthinite are relatively uncommon, and Fowey Consols is one of the better-known localities. This specimen (BM 34656) was bought from the dealer Richard Talling in 1862.

Bournonite, from the Herodsfoot mine, Liskeard, Cornwall (see fig. 61): bright grey multiply-twinned 'cogwheel' crystals, up to 5cm across, with tristetrahedral crystals (5mm) of tetrahedrite encrusted with chalcopyrite. The matrix is killas, with vein quartz which also forms thin septa. The association of bournonite and tetrahedrite on the same specimen is unusual.

In the years from about 1850 to 1875, sporadic cavities in the upper levels of the Herodsfoot lead mine produced bournonite crystals of a 'noble' quality unsurpassed elsewhere, either before or since. The best pieces were handled by the celebrated dealer Richard Talling; it is said that he was banned by the manager from entering the mine and from having dealings with the miners, but regained access by buying a few shares in the mine. Larger crystals are known from the locality, but as a group this specimen (BM 42223) is the most aesthetically pleasing; 22 × 16 × 12cm overall, it was bought from Talling in 1868 and bears his no. 3651.

Bournonite, from Wheal Boys, St Endellion, Cornwall (see fig. 61): an unusual cruciform twin (12mm) with irregular surface indentations, with sphalerite on fibrous massive jamesonite. The specimen (14cm overall length) is notable as one of the two originally described, as 'ore of antimony', by Philip Rashleigh in his *Specimens of British Minerals* (1797, p.34, Plate XIX); both of the figured specimens are now in the British Museum (Natural History), this one (BM 1964R,955) in the Russell collection.

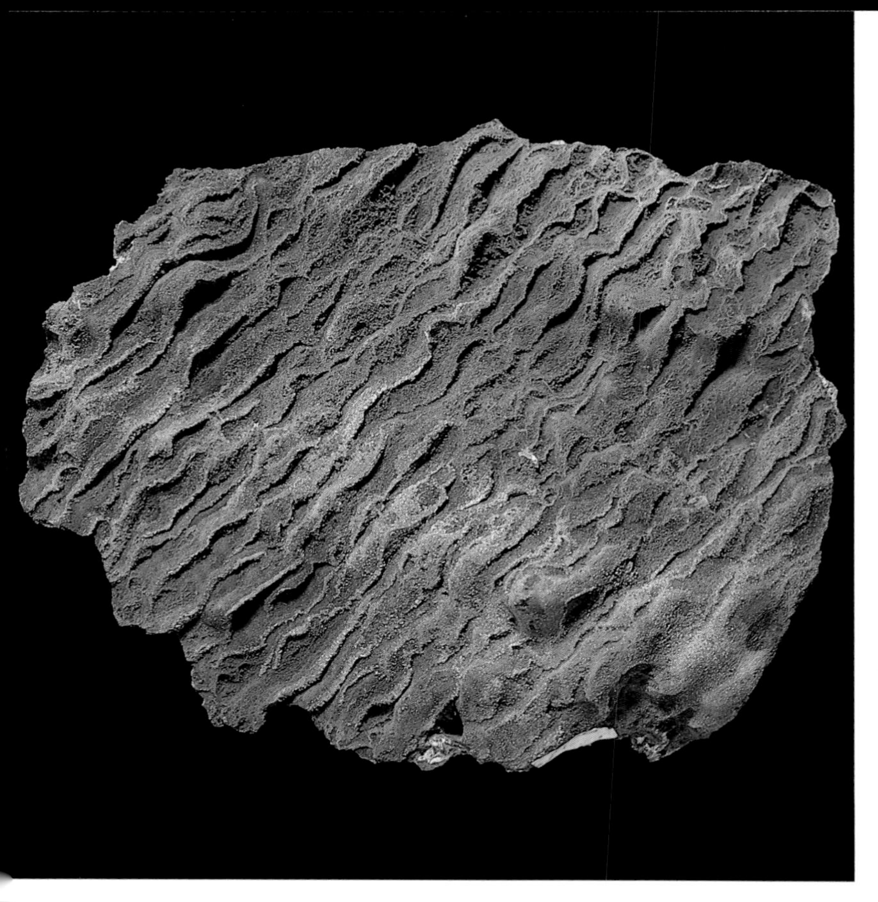

Brochantite, from the Fowey Consols mine, St Blazey, Cornwall (see fig. 61): a rippled green mass of minute crystals, with similar (but blue-green) langite, encrusting 'killas' (the Cornish term for clay-slate, the host rock at the mine). The two basic copper sulphates appear to have crystalized from mine waters flowing through a wide fissure in the slate.
The specimen (BM 36359) is about 14cm long, and was bought from the dealer Richard Talling in 1864.

Calcite, from Wheal Wrey, Liskeard, Cornwall (see fig. 61): L-shaped twins of acute rhombohedra (not prisms), about 5cm long, on a groundmass of much smaller calcite prisms encrusting galena.
This magnificent specimen (BM 32891) was bought from the dealer Richard Talling in 1861. It illustrates a general rule, that twinned crystals tend to be larger than accompanying single crystals of the same mineral. For a species that is so very common elsewhere, crystallized calcite is curiously rare in Cornwall.

Cassiterite, from the Prideaux Wood mine, Luxulyan, Cornwall (see fig. 61): brown zoned radiating fibres of 'wood tin', in altered granite. This obviously was once part of a spherule, 3cm in radius, which broke up to be later entrapped by its present matrix. This specimen (BM 32880) was bought from Mr Richard Talling in 1861, and bears his no. 629. Prideaux Wood was a small copper mine, abandoned and restarted several times, which also produced a little tin.

Cassiterite, from the Dolcoath mine, Camborne, Cornwall (see fig. 46): brown tapering crystals, the largest about 15mm long and with smaller crystals in twinned orientation, partly encrusted by light-brown siderite, and accompanied by dark-green botryoidal chlorite, on a quartz-chlorite matrix. Crystals of this habit are much less common than stumpy prisms, and are rare outside the Camborne mining district. Sometimes called 'needle tin', they were known to the old miners as 'sparable tin' from their resemblance to hand-made 'sparable nails', probably a corruption of 'sparrow-bill'. These curious crystals have somewhat rounded faces, and are thought to have formed at rather lower temperatures than the more equant prisms. This specimen (BM 44662) was bought from the dealer Richard Talling in 1872.
The Dolcoath mine was one of the best-known of the old Cornish mines, and had a long working life from early in the eighteenth century through to the twentieth century. Like many of the larger and richer mines, it incorporated several smaller mines and prospects, but was unusual in that it was worked profitably for both copper and tin, with arsenic as a by-product.

Cassiterite, from the Turnavore mine, St Agnes, Cornwall: a dark-brown, sharp, untwinned, equant crystal (8mm across, and one of several on the specimen), with quartz crystals on fine-grained altered slate. The large pyramid face {111} shows rounded growth features, and is bounded by narrow faces of {011}.
Turnavore was an ancient mine, later incorporated in the Wheal Kitty group (see fig. 57). This fine old specimen (BM 58223) was probably, but not certainly, in the Greville Collection, bought by the British Museum in 1810.

Cassiterite, var. 'wood tin', from the West Kitty mine, St Agnes, Cornwall (see fig. 57): a polished slice (9cm long) of a sea-worn pebble, showing repeated bands of light- and dark-brown fibrous cassiterite, with interstitial granular milky quartz and a little tourmaline.
Within the bands, the fibrous cassiterite is intergrown with quartz; the nature of the assemblage suggests that this characteristic variety of cassiterite formed from a gel, at lower temperatures than did larger crystals. Pebbles containing wood tin may still be found on the beaches near St Agnes. This one (BM 82746) was bought from the Redruth dealer William Peters in 1897.

Cerussite, from the Pentire Glaze mine, St Minver, Cornwall (see fig. 61): composite white prisms, up to 2·5cm long, on black manganese oxides coating quartz veinstone.
The Pentire Glaze mine, which ceased operation in 1857, was the most profitable of a small group working silver-lead veins which had been known since the sixteenth century. It is best known for its cerussite, which was recorded by Garby in 1848 (but not figured by Sowerby in 1817); crystals are reported to have reached a length of nine inches, on specimens no longer extant. Large specimens, but with much shorter crystals, are in the Truro Museum and the BM(NH). This comparatively modest specimen (BM 1964R,6853) is in the Russell collection; it was formerly in that of the collector John Hawkins, who acquired it in 1818, and is thus one of the oldest known.

Ceruléite, from Wheal Gorland, Gwennap, Cornwall (see fig. 61): a pale-blue botryoidal crust of minute fibres, in a cavity of brown 'gossany' quartz, with olivenite (not shown; the area in the picture is only about 4mm long). Ceruléite has not been recognized on any of the specimens from old Cornish collections, and may possibly have grown during the weathering of dump material.
Although the mine had long been abandoned, its dumps used to be among the few where one could be sure of finding interesting species. At the time of their removal for reprocessing, in the late 1970s, many specimens of ceruléite were uncovered. This specimen (BM 1957,455) was donated by Mr Arthur Kingsbury, who found it in 1948.

Chalcophyllite, from Wheal Unity, Gwennap, Cornwall (see fig. 61): sharp, green hexagonal plates, up to 5mm across, on dark-green chrysocolla and cornubite encrusting chalcopyrite, lining a cavity in brecciated quartz. Wheal Unity lay to the east of Wheal Gorland, on the same lode which yielded so many superlative specimens of copper and iron arsenates. Some of the chalcophyllite crystals on similar specimens are pale and cloudy, and are then found to be pseudomorphously altered to chrysocolla.
This specimen (BM 27091) was bought from Mr Richard Talling in 1856.

Chalcophyllite, from Wheal Gorland, Gwennap, Cornwall (see fig. 61): a group of composite crystals, the largest about 1cm across, with a little malachite. This specimen (BM 1964R,8988) is in the Russell collection, and the accompanying label reads:

> 'Philip Rashleigh MS catalogue Copper Ore No 633. Transparent bright green Hexagonal Micaceous Copper Ore with Malachites & Rich Copper Ore intermixed with Mundick from Tincroft Very rare & discovered 1792. This is the actual specimen figured in Philip Rashleigh's Specimens of British Minerals Part 1. 1797 Plate IX Fig 2. Though said to be from Tincroft, this specimen is almost certainly from Wheal Gorland. A.R.'

Chalcopyrite, from the New Kitty mine, St Agnes, Cornwall (see fig. 61): large sphenoidal crystals (up to 37mm edge), the twinning clearly shown by intersecting striations, with milky quartz prisms on altered slate.
This splendid specimen (BM 81730), 20cm long, was bought from Mr William Semmons in 1896; but it was probably mined many years earlier. Semmons was a London metal broker who had earlier worked as a clerk in the Cornish mines, and had amassed a large collection from which he later sold many of the best specimens.

Chalcopyrite, var. 'blistered copper ore', from the St Agnes district, Cornwall (see fig. 61): botryoidal mass, 14cm long. An unusual form for this species, it is like other botryoidal minerals in consisting of layers of fibrous crystals; sections (of other specimens) sometimes show interlayering by bornite. This specimen (BM 88584) was bought from the dealer Henry Heuland in 1829.

Chalcosine, from the St Ives Consols mines, Cornwall (see fig. 61): a detached group of lenticular, pseudo-hexagonal crystals, the largest about 12mm across.
Cornish chalcosine occurs in a variety of crystal habits, which have yet to be related to the probable conditions in which they grew. It is possible that different habits denote other related species, such as djurleite. This specimen (BM 34722) was bought from the dealer Thomas Renfree in 1862.

Chalcosine, from the Levant mine, St Just, Cornwall (see fig. 40): a fine twinned crystal, 4·5cm long, with smaller crystals and an intergrowth of white plates of the 'schiefer spar' variety of calcite. The surface of the crystal is somewhat 'sooty' and soils the fingers; the reason why this should be a characteristic property of many specimens of chalcosine, but by no means all, has not been explained. This specimen (BM 1905,207) was bought from Mr William Semmons.
The Levant mine is famous and remarkable for being one of the earliest to have its workings under the sea, and its first operation (under other names) dates from the eighteenth century. It was a rich copper mine, and also produced tin for which it is still worked as part of the Geevor mine. Picturesquely situated on the cliffs, its old steam winding engine has been preserved and is one of the historical exhibits to be seen by the visitor.

Childrenite, from the George and Charlotte mine, near Tavistock, Devon (see fig. 64): a fine brown crystal (12mm), one of the largest known from the locality, striated and showing parallel growth, with small cubes of pyrite, on dark-green chlorite encrusting slate.
The George and Charlotte mine was a small member of the Devon and Cornwall United group of copper mines, close to the banks of the River Tamar. This specimen (BM 96873) was bought with the Allan-Greg collection in 1860; the catalogue entry, in R. P. Greg's writing, states that he acquired it in 1856 from a Mr H. Bullen.

Clinoclase, from St Day, Gwennap, Cornwall (see fig. 57): radiating groups of slender laths, the terminations forming a blue-black drusy outer surface, accompanied by grey-green tufts of fibrous olivenite, on a matrix of altered granite 9cm long.
This specimen (BM 71875) was formerly in the collection of the Williams family, and was donated by Mr J. C. Williams in 1893. It undoubtedly came from the famous nearby mine, Wheal Gorland, from which clinoclase was described (as 'cupreous arsenate of iron') in 1801.

Connellite, from Wheal Muttrell, Gwennap, Cornwall (see fig. 57): a velvety crust of blue needles (2–3mm), with cuprite in small crystals and earthy ('tile ore'), lining a cavity in vein quartz.
Wheal Muttrell was a small mine, incorporated in Wheal Gorland before 1810. The Muttrell lode was the source of many of the finest old Cornish specimens of crystallized arsenates. Connellite was not recognized as a sulphate until 1850, when it received its present name. This old specimen (BM 1912,77) was bought from the dealer Samuel Henson in 1912, who had acquired it as part of the Isaac Walker collection; Walker, a wealthy brewer, originally bought it from the dealer Henry Heuland in 1832, as 'acicular light blue arseniate of copper, the velvet ore of Werner, select and most rare'; and it had probably passed through other hands before that.

Connellite, from Wheal Gorland, Gwennap, Cornwall (see fig. 57): azure-blue needles, up to 4mm long, in a cavity of massive cuprite ore. Other associated species (not shown) on this specimen are vein quartz, fine-grained cornubite, and platy chalcophyllite.
Wheal Gorland produced its first fine specimens of secondary copper minerals in the 1790s. Described by him as 'hair crystals of copper ore', this specimen was acquired by Philip Rashleigh at around that time; the species was not fully described until 1850. It is now part of the Russell collection (BM 1964R,12156).

Cuprite, var. chalcotrichite, from the Fowey Consols mines, St Blazey, Cornwall (see fig. 61): one of several nests of fine capillary crystals, in cavities of brecciated vein quartz partly encrusted with light- and dark-brown goethite.
This specimen (BM 363311), 10cm overall, was bought from the dealer Richard Talling in 1864.

Cuprite, from the Phoenix mine, Linkinhorne, Cornwall (see fig. 61): an unusually large (4cm), elongated cubo-octahedral crystal, with dark-brown goethite on rusty vein-quartz.
The Phoenix mine had its main period of activity in the middle half of the 1800s, and produced many fine specimens of crystallized cuprite and secondary copper phosphates. Most of the specimens were handled by the dealer Richard Talling, from whom this specimen (BM 44644) was bought in 1872.

Cuprite, from the Phoenix mine, Linkinhorne, Cornwall (see fig. 61): brilliant red, sharp cubo-octahedral crystals, up to 3mm, with brown limonite in a cavity in vein quartz, on killas. The edges of the cubo-octahedra are beautifully modified, by faces of the dodecahedron {110} and eicositetrahedron {211}. This specimen (BM 44193) was bought from Mr Richard Talling in 1871, bearing his no. 4001, and is yet another splendid memorial to his contributions to the preservation of Cornish minerals.

Erythrite, from the Pednandrea mine, Redruth, Cornwall (see fig. 57): bright needles (3–4mm long), on the surface of a fissure in quartzose slate, accompanied by fine-grained silvery 'smaltite'.
Cobalt ores only occurred in small quantities in Cornwall, and the erythrite resulting from their weathering is usually in the form of 'cobalt bloom' and but rarely well crystallized. This specimen (BM 1929,2) was bought from the London dealer Mr G. H. Richards in 1929.

Fluorite, from the Trevaunance mine, St Agnes, Cornwall (see fig. 57): purple composite crystals, 8mm edge and less, with tetrahedra/octahedra of black-brown sphalerite, and small prisms of quartz. The matrix is an earlier generation of sphalerite, with the crystal outlines visible under a coating of chlorite. The dominant form on the fluorite is the tetrahexahedron {310}. This specimen (BM 28317) was bought from the dealer Richard Talling in 1859, and bears his ticket number 147.
The Trevaunance mine, now long abandoned, was very profitably worked for tin from the middle of the eighteenth century. One of many mines in the St Agnes area, it was incorporated with them as parts of the Polberro mine in the nineteenth century and later as St Agnes Consols.

Fluorite, from the Trevaunance mine, St Agnes, Cornwall (see fig. 57): a group of pale purple crystals (10mm) with light-green centres, in parallel position and accompanied by prisms of colourless quartz. They are disposed on dark-green chlorite encrusting a matrix of altered 'killas', the Cornish term for clay slate. This specimen (BM 57754) is probably from the Greville collection, acquired in 1810.
The dominant faces on these fluorite crystals belong to the form {310}, the tetrahexahedron or 'four-faced cube'; the prominence of this is especially characteristic of Trevaunance, and of some neighbouring mines in the Polberro group. These mines were first worked nearly two centuries ago, but are long-since abandoned; a specimen from the Pell mine is figured in Rashleigh's *Specimens of British Minerals* (1797, Plate XXIII, Fig. 2).

▲ **Fluorite,** from St Agnes, Cornwall (see fig. 57): a 3cm composite crystal and smaller individuals, with quartz, black sphalerite, and pyrite, all disposed on chlorite encrusting 'killas'. The large fluorite has the shape of a rough cubo-dodecahedron, but the faces of its component crystals are combinations of the cube and the 'four-faced cube', with the purple colouring concentrated along the cube edges. The crystal habit was typical of the Trevaunance and neighbouring mines.

This specimen (BM 1964R,1103), part of the Russell Collection, has an interesting history. It was formerly part of the Walker Collection, which was formed in the early nineteenth century and acquired in 1912 by the London dealer Samuel Henson. Isaac Walker was a wealthy brewer, and his characteristic label (top line) shows that he bought it as lot no. 385 in the 1834 [April 15] auction sale of specimens belonging to the dealer Henry Heuland; the price paid is coded 'ooo', i.e. £3. ▶

Fluorite, from the Trevaunance mine, St Agnes, Cornwall (see fig. 57): purple crystals, on a finely-botryoidal crust of chlorite on 'killas'. The brightest individual in the picture is 7mm along the diagonal, and shows a small cube face at the apex of the {310} pyramid; nearby crystals show brightly-reflecting strip faces of the less-common dodecahedron.

This specimen (BM 1964R,1105) is in the Russell Collection.

◀ **Fluorite,** from Redruth (probably Carn Brea mine), Cornwall (see fig. 46): an almost-colourless cube with a pale purple outer zone, perched on and partly enclosing the tip of a matt crystal of milky quartz. Associated minerals are some smaller crystals of fluorite, of similar habit, bisphenoidal crystals of chalcopyrite, and a little specular hematite.

The edges of the cube, which is about 18mm across, are truncated by narrow faces of the rhombic dodecahedron {110}. This specimen (BM 52332) was bought in 1879 from the celebrated dealer, Richard Talling, and bears his numbered 'ticket' [4316].

▲
Fluorite, from the Carn Brea mine, Redruth, Cornwall (see fig. 46): an incomplete isolated cube, 6cm edge, associated only with a little micaceous hematite. Pale green and slightly cloudy at the core, clear at the corners, it shows to perfection an earlier mosaic assemblage and the smooth faces of the final stages of growth; the edges of the cube are modified by narrow faces of the form {310}.
This specimen (BM 70387) was bought in 1893 from Mr William Semmons, who at one time had worked for the Williams family of mine owners.

H — 1834 — 385 1103
Hexatetrahedral ame-
=thystine fluor with te-
=trahedral blende & rock
crystals St Agnes
000=£3 Cornwall

Fluorite, from Colcerrow quarry, Luxulyan, Cornwall (see fig. 61): an unusual crystal, deep-purple and 13mm across, with smooth octahedral and rough cube and dodecahedral faces, associated with quartz, apatite, and 'gilbertite' hydromica. The granite in the Colcerrow quarry, long abandoned, contained many vugs; some of these yielded apatite crystals of remarkable perfection.
This specimen (BM 1964R,1239) is in the Russell Collection, and was obtained from Mr William Semmons. Sir Arthur Russell was particularly fond of fluorite, and the superb suite of this species forms one of the many outstanding features of his collection.

Galena, from the Herodsfoot mine, Lanreath, Cornwall (see fig. 61): tarnished, tabular grey crystals of hexagonal appearance, up to 1.5cm across, with much smaller equant crystals and quartz. Although a lead mine, Herodsfoot produced few galena specimens of cabinet quality. The tabular habit is usually associated with twinning, and although it has not been observed elsewhere in Britain it is well known from several localities in Rheinland-Westphalia. This specimen (BM 42141) was bought from Mr Richard Talling in 1868, and was probably recently mined.

Galena, from Wheal Hope, Perranzabuloe, Cornwall (see fig. 57): prismatic pseudo-crystals, up to 8mm across, with a very little quartz on massive galena. Wheal Hope is the typical locality in Cornwall for these pseudomorphs of galena after pyromorphite, often known as 'blue lead', but there are several other localities for them in the world.
This specimen (BM 81971) was bought from Mr William Semmons in 1897, without a locality, but it is sufficiently characteristic for there to be very little doubt that it is from Wheal Hope.

Goethite, from the Restormel mine, Lanlivery, Cornwall (see fig. 61): stout black composite prisms, 5cm long, and smaller single prisms, with a little quartz showing two generations of growth, on massive goethite. Goethite crystals of this size and quality are exceptional in their own right, but Greg & Lettsom (1858) state that some dealers used to pass them off as rutile after heat treatment.
This specimen (BM 26820) is the finest known from the locality, and was bought from Mr Richard Talling (who lived in the nearby town of Lostwithiel) in 1855.

Goethite, from the Restormel mine, Lanlivery, Cornwall (see fig. 61): radiating fibres, 4·5cm long, with strong concentric zoning in shades of brown, with a little quartz. The complete specimen (BM 26586), 18 × 15 × 14cm, was bought from Mr Richard Talling in 1850.
The Restormel mine, situated to the south-west of the Bodmin Moor granite mass, was Cornwall's most productive iron mine and was visited by Queen Victoria in 1846. The main lode, running almost north-south, was mined for a distance of two miles and traced for five miles; the mineralization took place in many episodes, and Collins (1912) observed thirty-six quartz bands across its width, representing at least eighteen widening movements.

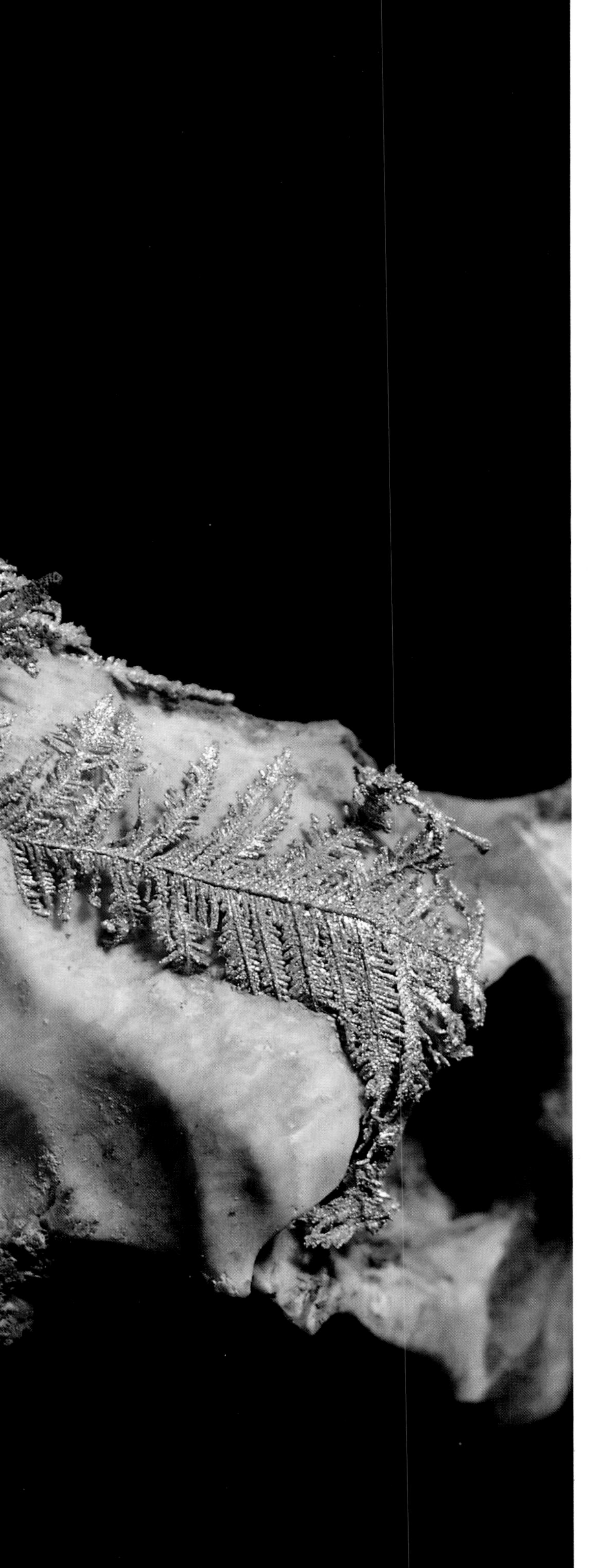

Gold, from Hope's Nose, near Torquay, Devon (see fig. 64): a beautifully delicate dendritic growth in cream-coloured calcite, with brown weathered dolomite. Originally wholly enclosed by calcite, the gold has been exposed by acid treatment.
The small veins at this locality outcrop on the sea coast, near the sewage outfall of Torquay, and are remarkable for the palladium minerals isomertieite and mertieite-II which have recently been found there in small amount. This specimen (BM 1981,458) was collected by Sir Arthur Russell; it showed little visible gold, and was regarded as a low-quality duplicate until subjected to acid treatment in the course of the palladium study (figured area 3.5 × 2.8cm).

◀ **Libethenite,** from the Phoenix mine, Linkinhorne, Cornwall (see fig. 61): dark-green composite crystals (up to 12mm), on yellowish-brown botryoidal dufrenite encrusting banded vein quartz.
These remarkable crystals of libethenite are from the Stowe's section of the Phoenix mine, which produced many other interesting phosphates including andrewsite and fine specimens of chalcosiderite. The related species rockbridgeite and kidwellite have been found with the dufrenite. This specimen (BM 61415) was bought from the London dealer Dr F. H. Butler in 1886.

Liroconite, from Cornwall: a group of blue crystals, up to 18mm across, of the characteristic habit resembling a flattened octahedron, on milky quartz. Although it has no closer localization, there can be no doubt that this is from the celebrated oxidized lode running through Wheal Gorland and Wheal Unity, in Gwennap parish, which produced so many classic specimens of copper arsenates (see fig. 57).
This specimen (BM 1951, 427) was formerly in the collection of Professor Nevil Story-Maskelyne, and was given to the BM(NH) in 1951 by a member of his family, Mr J. Arnold-Foster.

Liroconite, from Wheal Muttrell, Gwennap, Cornwall (see fig. 57): deep-blue obtuse pyramidal crystals, about 2.5mm across, with encrusted and intergrown needles of olivenite, on limonite and quartz gossan.
This specimen is no. 827 in the Rashleigh collection, at the County Museum in Truro, and is figured in Rashleigh's 'Specimens of British Minerals' 1802, Part II, Plate XI, Fig. 3. The entry in the MS catalogue reads: '827 Copper Ore of a Sky blue colour with double four sided pyramids a little transparent and very perfect, upon needle xtls of Copper Ore of a Grass green colour and opake, upon an ochry Stone of copper ore, r.r.r.r. [i.e. exceedingly rare] Huel Mutterel'.
▼

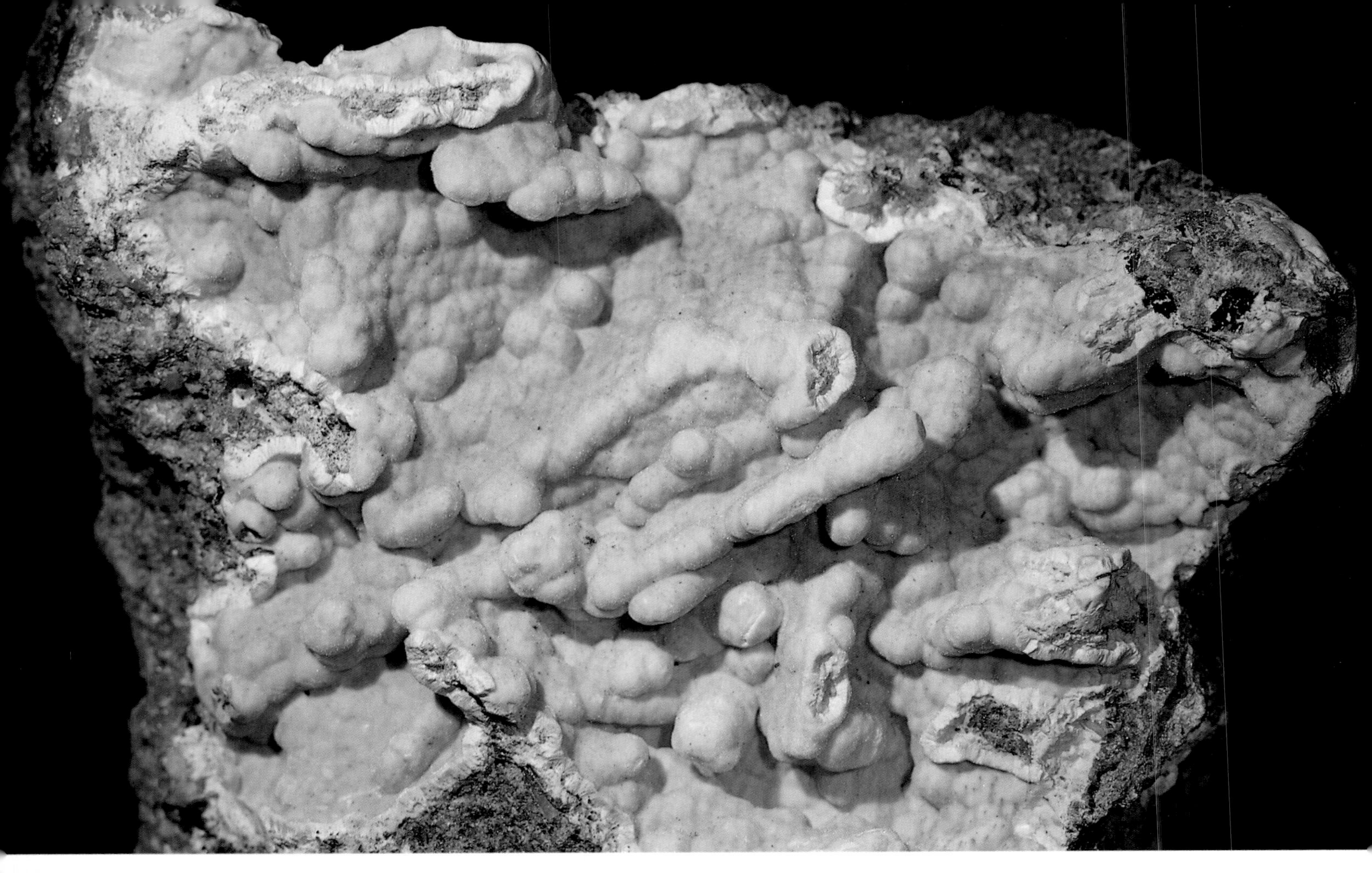

Liskeardite, from the Marke Valley mine, Liskeard, Cornwall (see fig. 61): soft fibrous crystals, forming a pale-green botryoidal crust on a fine-grained matrix of chlorite and quartz with a little chalcopyrite. An hydrated basic arsenate of aluminium and iron, liskeardite is poorly characterized and its crystals have as yet proved unsuited to X-ray structural analysis.
This specimen (BM 50821) was bought in 1877 from Mr A. K. Barnett of Penzance, a lecturer in mining and part-time mineral dealer, and is one of those examined by Professor Maskelyne when he first described liskeardite in the following year.

Ludlamite, from Wheal Jane, Kea, near Truro, Cornwall (see fig. 57): sharp but composite green crystals, the foreground sheaf 33mm tip to tip, with a little limonite, on a matrix of drusy siderite underlain by pyrite.
Wheal Jane was already an old mine when it was reopened for the second or third time in 1851, and during this period of working ludlamite was first discovered and described in 1877. After a long period of disuse, the mine has been restarted (for tin) twice in the past dozen years and is one of the few remaining active Cornish mines. Some excellent specimens of ludlamite have been recovered recently; this one (BM 1985,679) was acquired by exchange from the Lancashire dealer, Mr Ralph Sutcliffe.

Mimetite, from Wheal Unity, Gwennap, Cornwall (see fig. 57): pale brown prisms, up to 4mm dia., with quartz crystals and a little chlorite on altered granite. Wheal Unity was a rich copper mine, immediately to the east of and working some of the same veins as the better-known Wheal Gorland; the mimetite was found at a depth of about 50 fathoms from surface.
This specimen (BM AG.11) is from the famous Allan-Greg collection, which was bought by the British Museum in 1860. Thomas Allan, in the very detailed catalogue, records that it was 'analysed and given to me by the late Mr Gregor, who found it contained arsenical acid.' The Revd William Gregor, discoverer of titanium, published his analysis of 'native arseniate of lead' in 1809 and recorded chlorine as an essential constituent. Gregor did not keep a collection, and this is one of the very few specimens known to have passed through his hands.

Olivenite, from Wheal Unity, Gwennap, Cornwall (see fig. 57): dark green prisms, about 2mm long, with a little quartz and carthy hematite. The complete specimen is about 12cm long, and consists of rough crystals of milky quartz almost wholly covered by a reniform crust of olivenite needles.
This specimen (BM 27078), bought in 1856, is an example of those mentioned by Greg & Lettsom (1858): 'Mr Richard Talling has lately obtained excellent specimens at the old workings of the Wheal Unity, near St Day.' Wheal Unity was an old copper mine adjacent to Wheal Gorland (see caption to 'Mimetite').

426

▲ **Olivenite,** from Wheal Jewel, Gwennap, Cornwall (see fig. 57): bands of radiating fibres, showing concentric colour zoning from buff to dark brownish-green, touching massive malachite and enclosing a little cuprite, on quartzose matrix.
The exact locality is in doubt, being Huel Jewel in Rashleigh's ms catalogue and Huel [Wheal] Providence in the (later) caption to his published figure. The former seems more likely, in view of the abundant occurrence of olivenite in Gwennap parish; if the latter, the only candidate seems to be a small copper mine near Carn Brea, also known as South Wheal Tincroft. At different times, there were at least eight Cornish mines with the name Wheal Providence!
This specimen is figured in Rashleigh's 'Specimens of British Minerals' 1802, Part II, Plate IX, Fig. 5, and is no. 969 in the Rashleigh collection at the County Museum, Truro.

Orthoclase, from the Brill No. 1 quarry, Constantine, Cornwall (see fig. 57): a pink crystal of 'Baveno' habit, 15cm long, with other individuals in parallel (untwinned) position at one end and a terminated quartz prism attached to one side.
This locality is one of several quarries at the south edge of the Carnmenellis granite mass, most of which have revealed pegmatitic cavities at some stage of their operation. This specimen (BM 1964R,10676) is from the collection of the late Sir Arthur Russell, who found it in 1948.

◀ **Olivenite,** from Wheal Muttrell, Gwennap, Cornwall (see fig. 57): radially-streaked concentric bands of fibres round a glassy brown central core, with colour zoning in shades of buff, on quartz and massive impure djurleite.
Fibrous varieties of olivenite were common in Gwennap, and were known to the miners as 'wood copper'. The name olivenite conveys the idea of green, but when the mineral is fibrous the colour can vary, from pure white through various shades of green and buff to dark brown. This variability of colour seems to be independent of the size of the fibres, and although noted nearly two centuries ago it has yet to be explained.
This specimen, 4 × 3 × 3cm, is figured in Rashleigh's 'Specimens of British Minerals' 1802, Part II, Plate XI, Fig.1, and is no. 726 in the Rashleigh collection at the County Museum, Truro.

▲
Psilomelane, from the Monkstone manganese mine, Brent Tor, Devon (see fig. 64): a bright botryoidal mass, about 7cm long. This specimen (BM 1964R,3390) is in the Russell collection; it was bought in 1911 as part (no. 709) of the collection of the Tavistock surgeon Edmund Pearse (1788–1856), which was better known for the fine suite of minerals from the Virtuous Lady mine which it contained.

Pyrite, from the Virtuous Lady mine, Buckland Monachorum, Devon (see fig. 64): fine grained, intergrown with some siderite and forming a tapering cast 26cm long × 9·5cm at its widest part, the original crystal around which it grew having been dissolved by some natural process. The part of the specimen on the right is an incomplete cast of the same type, and not a fragmented 'box'.
This is perhaps the largest known example of a 'lady's slipper' epimorph, usually composed of siderite, which the Virtuous Lady mine yielded in quantity in 1832-1833 and for which it is almost the only locality. Greg & Lettsom, in their *Mineralogy* (1858:260), thought that the original crystals must have been gypsum; but, following Russell, it is now generally supposed that these casts are after baryte, of an unfamiliar habit.
This specimen (BM 1964R,12342) is part of the Russell collection, and was formerly in that of the Tavistock surgeon Edmund Pearse (1788–1856), acquired by Sir Arthur in 1911.
▼

Pyromorphite, from Wheal Alfred, Phillack, Cornwall (see fig. 40): tapering yellowish-green crystals, some hollow and up to 4mm across, on iron-stained quartz. Pyromorphite specimens from this locality are the finest found in Cornwall, but it is not known from which part of the mine they came.

On the site of a much older tin mine, Wheal Alfred enjoyed a short spell of prosperity as a copper mine in the first quarter of the nineteenth century. A description (Phillips, 1814) and sketch plan show that several veins intersected, and the mineralization must have been complex for the records of output show that lead and tin were also produced. Another brief period of operation from 1851 to 1862, as Great Wheal Alfred, was ultimately unsuccessful.

This specimen (BM AG.46), part of the Allan-Greg collection bought by the British Museum in 1860, was given to Thomas Allan in 1826 by the famous mining engineer John Taylor, who owned the mine from 1823–1826.

Quartz, var. chalcedony, from the Trevascus mine, Gwinear, Cornwall (see fig. 40): a light brown, hollow branching stalactitic mass, about 16cm long. This specimen (BM 35898) was bought from Mr Richard Talling in 1864, but may well have been found much earlier.

Trevascus was an old mine, started for copper in the early years of the eighteenth century and later producing tin until the middle of the nineteenth century. It produced some galena, catalogued by Dr John Woodward in 1728. Mentioned by Borlase in 1758, it is best known for the beautiful and fantastic shapes assumed by some of its chalcedony specimens.

◀ **Quartz,** var. amethyst, from Wheal Uny, Redruth, Cornwall (see fig. 46): clear purple terminations on prisms of much paler, milky quartz, part of the lining (7cm overall) of a once-larger cavity in vein quartz in granite. Wheal Uny was a small copper mine.
Crystallized amethyst is rare in Cornwall, and this specimen (BM 39940) is exceptional; it was bought in 1866 from the Redruth dealer Mr William Peters.

Quartz, var. chalcedony, from the Haytor mine, Ilsington, Devon (see fig. 46): a 10cm group of iron-stained, well-defined pseudo-crystals after datolite.
Originally described in 1827 as a new mineral, haytorite, it was soon discovered that the crystals had the same angles as datolite. Until the principles of the pseudomorphic replacement of one mineral by another became better understood, the unusual perfection of many haytorites presented the early crystallographers with a difficult problem. Little is known of its mode of occurrence in the Haytor iron mine, and less of the earlier phase of mineralization that produced the original datolite crystals. Unaltered datolite from Devon or Cornwall has only been found as very much smaller crystals than these. This specimen (BM 71844) was once in the Williams collection, and was presented by Mr J. C. Williams in 1893.

▲
Quartz, from the Virtuous Lady mine, near Tavistock, Devon (see fig. 64): a group of milky, polysynthetic crystals, up to 8cm across, partly encrusted by dark-brown, rounded lenticular siderite. The prism faces are rough and tapering, whereas the terminations are smooth; there are thin layers of clay beneath these terminal faces, which on some specimens have caused the outer part of the crystal to lift away. In consequence, the trivial name 'capped quartz' has been given to this variety.
This specimen (BM 43560) was bought from the London dealer Bryce Wright in 1870. The writer John Ruskin, who was something of a quartz fanatic, included it in a special exhibit that he was once allowed to arrange at the BM(NH).

◀ **Scheelite,** from the Ramsley mine, Sticklepath, Devon (see fig. 64): iron-stained interpenetrating bipyramids, with a little pyrite, forming a 4·3cm group. This specimen (BM 1958,603) was donated by Mr Arthur Kingsbury, who found it on the old dumps in 1958.
The Ramsley mine was worked for copper at the beginning of the present century. Rough octahedra of fine-grained quartz were known from it for many years; but they were thought to be pseudomorphs after fluorite, until this specimen (and smaller ones showing partial alteration) showed that the original crystals were scheelite.

Siderite, from the Virtuous Lady mine, Buckland Monachorum, Devon (see fig. 64): a crust of rounded brown crystals, forming a hollow cast or epimorph 9cm across, containing prisms of milky quartz on aggregated chalcopyrite crystals. The original cubic crystal, around which this beautiful specimen grew, has completely disappeared; but the evidence that it was fluorite is provided by characteristic imprints of mosaic cubes on the inner surface. The chalcopyrite and quartz appear to have been formed after the fluorite crystal dissolved away, as first suggested by de la Beche in 1839.

Only a few of these 'box' epimorphs have been preserved, of which this is by far the best; but the mine is also noted for another shape of siderite epimorph, the 'lady's slipper'. Many of these have the outline of a gothic arch, and are thought to have formed around tapering baryte crystals. Found in considerable numbers in 1832–33, some were connected to the 'boxes' and many have pyrite intergrown with the siderite. The conditions under which fluorite and baryte could have been dissolved away, leaving siderite unchanged, have yet to be explained: 'The truth is, we are not yet in possession of a sufficient number of facts to warrant our coming to any conclusion on the subject.' (letter from Edmund Pearse, October 2, 1835, to Mrs Bray and published by her).

This specimen (BM 21338) was bought in 1847 from the Tavistock surgeon Edmund Pearse (1788–1856), whose collection was acquired by Sir Arthur Russell in 1911.

(see also Pyrite, p.118).

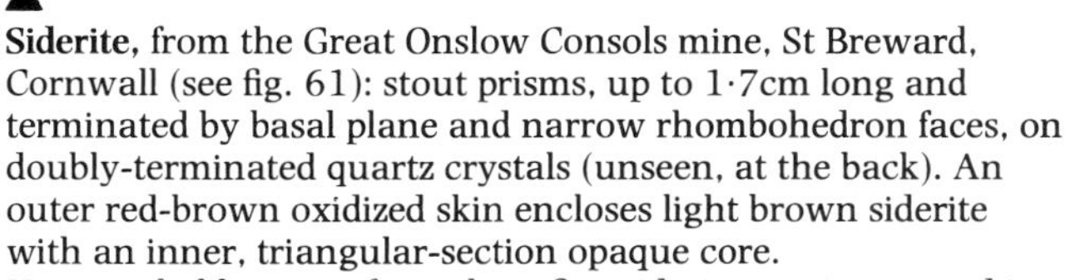

Siderite, from the Great Onslow Consols mine, St Breward, Cornwall (see fig. 61): stout prisms, up to 1·7cm long and terminated by basal plane and narrow rhombohedron faces, on doubly-terminated quartz crystals (unseen, at the back). An outer red-brown oxidized skin encloses light brown siderite with an inner, triangular-section opaque core.

Unremarkable, apart from these fine siderite specimens and its rather grandiose name, this mine produced relatively small amount of copper ore and a little pyrite in the mid-nineteenth century.

This specimen (BM 26752) was bought from Mr Richard Talling in 1854, and is one of the few known from the locality.

Siderite, from Wheal Maudlin, Lanlivery, Cornwall. Close up of short prismatic crystals with basal plane and showing concentric zoning. The majority of the crystals about 13mm across, and lining a geode (see fig. 76 for complete geode).

Spangolite, from Wheal Gorland, Gwennap, Cornwall (see fig. 57): dark blue, tapering hexagonal tabular crystals, up to 2mm across, part of a group 2cm across, on killas. This specimen (BM 1964R,5364) is in the Russell collection, and is one of only four or five known from this locality; all were formerly in the collection of the Williams mine-owning family, of Gwennap.
Spangolite has a micaceous basal cleavage, on which the colour has a greenish cast, and although a basic sulphate-chloride, it bears some superficial resemblance to the arsenate chalcophyllite. The Cornish material was first noted in 1893 by H. A. Miers, while he was making a selection of specimens from the Williams collection for the BM(NH), shortly after Penfield's original description in 1890. The evidence for Wheal Gorland (or, rather, its main lode) as the locality is circumstantial, but strong.

Stephanite, from Wheal Newton, Harrowbarrow, near Calstock, Cornwall (see fig. 64): a lustrous, black crystal, 37mm (diag.) × 18mm (thick), with smaller individuals irregularly disposed beside it, on crystallized iron-stained ankerite and siderite intergrown with and encrusting zoned milky quartz.
The crystal is exceptionally large for the species, and was described by Prof. W. J. Lewis. Smaller crystals were also found in Cornwall, at Wheal Ludcott; but the only other published provenance, Wheal Boys, is less certain because the matrix of the single known specimen, despite the labelling, is of untypical appearance for the locality.
This specimen, (BM 1985MI,4595), one of the treasures of the mineral collection in the Geological Museum (formerly the Museum of Practical Geology), was bought in 1880 from Captain H. Bennett of Redmoor [near Callington], Cornwall, who probably once worked at Wheal Newton.

Stibnite, from Wheal Boys, St Endellion, Cornwall (see fig. 61): an isolated group of radiating grey prisms, up to 1·5cm long. This specimen, 4 × 3 × 2cm, is in the Rashleigh collection at the County Museum, Truro.

Although Cornwall is noted for the excellence of its specimens of the antimony sulphosalts bournonite and tetrahedrite, good examples of Cornish stibnite are rare; there are none at the BM(NH).

Antimony ore, doubtless stibnite, was recorded in the St Endellion area from the Roscarrock estate, in an entry in Nehemiah Grew's 1681 catalogue of the Royal Society's Collection: 'STYRIATED ANTIMONY, also Native, from Cornwall; called ROSCARROCKS. A Congeries of strait, long, slender, and edged *Styriae*, of a bright Steel-colour, almost like a cluster of small broken *Needles*.' The specimen, unfortunately, has not survived.

Tourmaline, from Woolley Farm, near Bovey Tracey, Devon (see fig. 64): a composite, doubly-terminated crystal of schorlite, 9cm across and 8cm long, with a little limonite and (at the back) crystals of apatite. Specimens were first found at this classic locality a year or two before 1817.

This specimen (BM 1912,191) is the largest known from this classic locality, and was presented by Mr F. N. A. Fleischmann (later, F. N. Ashcroft) in 1912, who bought it from the dealer Samuel Henson as part of the Walker Collection. A characteristic, glued-on 'ticket' bearing the number 50 indicates that it was earlier in the private collection of Henry Heuland; the catalogue to this collection is lost, but a partial copy (made in 1873), lacking entry #50, shows that #51 was also tourmaline from Bovey Tracey.

Gold, from Hope's Nose, near Torquay, Devon; arborescent ▶ aggregates of cubo-octahedral crystals on cream coloured calcite. This specimen (BM 1985,539) about 5.2cm long was bought from the dealer Simon Harrison in 1985.

◀ **Turquoise,** from the West Phoenix mine, Linkinhorne, Cornwall (see fig. 61): blue-green spherules, up to 5mm dia. and with a crystallized surface, on dark-brown hisingerite encrusting a cavity in limonite gossan. Now recognized as the variety ferri-turquoise, this mineral was described as a new species in 1876 and named henwoodite in honour of the Cornish mining engineer W. J. Henwood. Its occurrence at the West Phoenix mine is closely similar to that of andrewsite and chalcosiderite. This specimen (BM 50139) was bought from Mr A. K. Barnett in 1876.

Wavellite, from High Down quarry, Filleigh, near South Molton, Devon (see fig. 61): a botryoidal slab of radiating crystals, pale green on one side of the centre and blackened by carbonaceous inclusions on the other. The largest sphere is 5cm across, and much bigger than the average 1cm or less for the locality; exceptionally, a little variscite has been found at the centre of the spheres on some specimens. It is more usual for the wavellite to be disposed on the surfaces of rhomboidal jointing in black slate, which is bleached near the mineralization. High Down quarry is the type locality for wavellite, named in 1805 for Dr William Wavell who is generally supposed to have first found it; but an earlier discovery before 1785 by a Mr I. Hill, is also claimed (Sowerby, 1811). Old reference works and labels often give the locality as Barnstaple, the nearest town of any size. At first thought to be a zeolite, from its appearance, and for a while called hydrargillite, wavellite gave problems to the early analysts because aluminium and phosphate were difficult to determine in each other's presence.

This specimen (BM 1922,1181) was donated by Mr W. H. Reeve, who found it in 1913.

▼

REFERENCES

References

(with annotations, and selective bibliography)

The list that follows contains more than 1100 entries, most – but not all – of which relate to the minerals, mines, or geology of Cornwall and Devon. The few exceptions are those concerned in some way with the chapter on collectors and dealers.

By far the most significant and numerous omissions are references to manuscript sources, and particularly to the many thousands of contemporary mining reports and accounts in local newspapers and the technical press. The first Cornish newspaper was *The Royal Cornwall Gazette* (1801-1951), soon followed by *The West Briton* (1810–) and later by *The Western Morning News* (1860–). Edited excerpts on various subjects (including mining) from *The West Briton* are contained in R. M. Barton's *Life in Cornwall* (1970–74). Earlier, in the eighteenth century, Cornwall and Devon were covered by *The Sherborne Mercury* (1737–) and *The Exeter Flying Post* (1763–). For many years, *The Mining Journal* (London) (1835–) contained too much Cornish material to be itemized in the indexes.

No complete catalogue or bibliography of works in our present field of interest exists, but we have had a reciprocal concern with that published by Halls *et al.* (1985). At the end of last century, Whitaker (1870, 1874) produced separate lists for Cornwall and for Devon, and lengthy general works were published by Boase & Courtney (1872–1882) and by Boase (1890). Extensive lists of references are features of the 'District Memoirs' and other monographs of the British Geological Survey, formerly entitled the Geological Survey of Great Britain and the Institute of Geological Sciences.

Geological Survey publications are collected together, at the end of the listing.

Topographical maps are produced and published by the Ordnance Survey, and may be obtained through booksellers or from the official retail agents, Messrs Cook, Hammond and Kell, Ltd, 22 Caxton Street, London SW1. The most common series have scales of 1:250 000, 1:50 000, and 1:25 000 (the first two replace the earlier 1/4 inch to 1 mile, and 1 inch to 1 mile series, respectively; the last, or 'Pathfinder' series, had no predecessor). Maps to the larger scale of 1:10 000 are replacing the earlier 6 inches to 1 mile series, and are recommended for serious field work.

Abbreviations used: The abbreviated titles of serial publications should be self-explanatory; the exception, *TRGSC*, for Transactions of the Royal Geological Society of Cornwall, is used to save space (there are 125 references to it). Other abbreviations include 'n.d.' (no date); 'n.s.' (new series); and 'ppb.' (paperback).

ABBOTT, G. (Jun.) 1833 *An Essay on the Mines of England*; their importance as a source of national wealth, [etc.]. London. (esp. ch.vii, pp.77–92, on Cornish copper mines)

ABBOTT, H. A. 1920 The Levant Disaster. *Mining Mag.* **22**:207.

AHMAD, S. N. 1977 The geochemical distribution and source of copper in the metalliferous mining region of southwest England. *Mineralium Deposita* **12**:1–21

AIKIN, A. *see* Rashleigh

ALDERMAN, A. R. 1935 Almandine from Botallack, Cornwall. *Mineral. Mag.* **24**:42–48

ALDERTON, D. H. M. 1975 Fluid inclusion studies in SW England. *Proc. Ussher Soc.* **3**:214–217

—— & JACKSON, N. J. 1974. The distribution of tin in the metabasite hornfelses within the St. Just metamorphic aureole. In: Stemprock, M. (ed.) *Metallization associated with acid magmatism*. Geological Survey, Prague. **1**:297–303

—— & —— 1978 Discordant calc-silicate bodies from the St. Just aureole, Cornwall. *Mineral. Mag.* **42**:427–434

—— & RANKIN, A. H. 1983 The character and evolution of hydrothermal fluids associated with the kaolinised St Austell granite, SW England. *J. Geol. Soc. London* **140**:297–310

ALLAN, R. 1834 *A Manual of Mineralogy Comprehending The More Recent Discoveries in The Mineral Kingdom*. Edinburgh.

ALLMAN-WARD, P., HALLS, C., RANKIN, A. H. & BRISTOW, C. M. 1982 An intrusive hydrothermal breccia body at Wheal Remfry in the western part of the St Austell granite pluton, Cornwall. In: Evans (ed.), pp.1–28

ALLOM, T. 1831 *see* Britton & Brayley 1831

ANON. 1671 An Accompt of some Mineral Observations touching the Mines of Cornwal and Devon; wherein is described the Art of Trayning a Load; the Art and Manner of Digging the Ore; and the Way of Dressing and Blowing Tin: Communicated by an Inquisitive person, that was much conversed in those Mines. *Phil. Trans. R. Soc. London* **6**:2096–2113

ANON. 1790 Auszug aus dem Reisejournal eines Deutschen. / Reise von Londen in die Grafschaft Kornwall. *Bergmännisches J.*, Jahrgang 3 (Band 2), pp.1–40, 143–170. (journey made in 1783).

ANON. 1817 *The gazetteer of the county of Cornwall*, to which is prefixed a survey of the county. John Heard, Truro. (contains a section of 23 unnumbered pages on minerals and mines)

ANON. 1832 On the mining district of Redruth. *Q. Mining Rev.* **2**:201–225

ANON. [1854] The Mines and Miners of Cornwall. pp.1–32, offprint from *Chambers's Repository* (n.d., ?1854)

ANON. 1862 The Cornish Man-Engine. *Mining & Smelting Mag.* **1**:366–384

ANON. 1864 The System of Selling Tin-Ore in Cornwall. *Mining & Smelting Mag.* **5**:6–11, 132–135 (with letters from the *West Briton*)

ANON. 1864 *Report of the Commissioners appointed to inquire into the condition of all mines in Great Britain*. HMSO, London.

ANON. 1880 *Sketch of the Life of William West, C.E., of Tredenham*. Brendon, Plymouth. (ppb. reprint, Inst. Cornish Studies, Redruth, 1973; West was a successful steam engineer in Cornwall).

ANON. 1892 Excursion to Caerhays Castle. *Trans. Mining. Assn Inst. Cornwall* **3**:207–213 (p.213 footnote Williams coll.).

ANON. 1898 Cornish Diamonds [var. anecdotes]. *Cornish Mag.* **1**:486

ANON. 1909 Particulars of Cornish Mines. In: *Report of the County Commissioners of the Cornwall Mining Exhibit, Imperial International Exhibition, Shepherd's Bush, London.*

ANON. 1953 Cornish Engines Preservation Society / Report of the Survey Sub-committee. 2nd revised and illustrated ed., 1953 (1st ed. 1943; revised ed. 1950).

ANON. 1974 *Mining in Cornwall Today*. Cornish Chamber of Mines, Truro.

ANTHONY, G. H. 1968 *The Hayle, West Cornwall and Helston Railways*. Oakwood Press, Lingfield.

ARGALL, W. H. 1872 On Gossans. *Rep. Miners' Assn Cornwall & Devon* (1871–72), pp.37–41

—— 1874 On the occurrence of wood-tin ore in the Wheal Metal lode at Wheal Vor in Breage. *J. R. Inst. Cornwall* **4**:255–256

—— 1876 On the elvan courses of Cornwall. *Rep. Miners' Assn Cornwall & Devon* (1875), pp.37–64

ASHCROFT-HAWLEY, V. R. G. & MITCHELL, D. 1960 *Devon Ball Clays and China Clays*. Watts, Blake, Bearne & Co. Ltd. Newton Abbott.

ATKINSON, B. 1983–1984 A survey of the lesser known mining sites of Cornwall. List 1 – The Liskeard area. *J. Plymouth Mineral & Mining Club* **13**(3):14–15; List 2 – Calstock, Callington and Launceston. **14**(1):17; List 3 – St Austell to Saltash. **14**(2):13; List 4 – Around St Ives. **14**(3):16; List 5 – St Agnes to Perranporth. **15**(1):9

ATKINSON, M. (with Burt, R. & Waite, P.) 1978 *Dartmoor Mines / the mines of the granite mass*. Exeter.

ATKINSON, R. L. [1985] *Tin and Tin Mining*. (No.139, Shire Album series). Shire Publications, Princes Risborough.

BABINGTON, W. 1796 *A Systematic Arrangement of Minerals*, [etc.]. 2nd ed., London. (1st ed. 1795)

—— 1799 *A New System of Mineralogy, in the form of a Catalogue*, [etc.]. London. (dedicated to Sir John St Aubyn, who had by then bought Babington's collection – founded on that of the Earl of Bute. Now in the Plymouth City Museum;- see Curry 1975, Torrens 1977)

—— *see* Davy 1805

BADHAM, J. P. N., STANWORTH, C. W. & LINDSAY, R. P. 1967 Post-emplacement events in the Cornubian batholith. *Econ. Geol.* **71**:534–539

—— & —— 1976 The curved-crystal pegmatite, Goonbarrow. *Proc. Ussher Soc.* **3**:411–447

—— 1980 Late magmatic phenomena in the Cornish batholith – useful field guides for tin mineralization. *Proc. Ussher Soc.* **5**:44–53

—— 1982 Strike-slip orogens – an explanation for the Hercynides. *J. Geol. Soc. London* **139**:493–504

BAKER, O. 1982 Lavin's Museum, Penzance. *The Geological Curator* (new title of *Geol. Curators' Group Newsletter*) **3**(5):316–318.

BALCHIN, W. G. V. 1954 *Cornwall, an illustrated essay on the history of the landscape*. London.

BALLANTYNE, R. M. 1869 *Deep Down / A Tale of the Cornish Mines*. Nisbet, London. (later eds. 1912, 1913)

BALL, T. K. & BASHAM, I. R. 1979 Radioactive accessory minerals in granites from south-west England. *Proc. Ussher Soc.* **4**:437–448

——, —, & MICHIE, U. McL. 1982 Uraniferous vein occurrences of South-west England; paragenesis and genesis. In: *Vein-type and similar uranium deposits in rocks younger than Proterozoic*; Proceedings of a technical committee meeting, Lisbon 24–28 Sept. 1979. Vienna IAEA. pp.113–158

——, — ,BLAND, D., & SMITH, T. K. 1982 Aspects of the geochemistry of bismuth in south-west England. *Proc. Ussher Soc.* **5**:376–382

BANCROFT, P. 1973 *The World's Finest Minerals and Crystals*. Viking Press, New York. (cuprite p.54, chalcocite p.100, siderite p.166)

—— 1984 *Gem and Crystal Treasures*. Western Enterprises, Fallbrook, and Min. Record, Tucson. (ch.81, Herodsfoot mine bournonite, also other mine and mineral photos, pp.377–380)

BANNISTER, F. A., HEY, M. H., & CLARINGBULL, G. F. 1950 Connellite, buttgenbachite, and tallingite. *Mineral. Mag.* **29**:280–286

BARCLAY, C. F. 1931 Some Notes on the West Devon Mining District. *TRGSC* **16**:157–176

BARING-GOULD, S. 1899 *A Book of Cornwall*. Methuen, London. (2nd edn, 1902; 3rd edn, 1912).

BARNETT, A. K. 1869 A ramble through Wheal Buller, West Basset and South Frances Mines. *Rep. Miners' Assn Cornwall & Devon* (1869), pp.33–35

—— 1870 Mineral Phenomena (Gwennap). *Rep. R. Cornwall Poly. Soc.* **37**:53–56

—— 1874 Observations on the elvan courses, greenstones and sandstones of Cornwall, with remarks on their associated minerals. *Rep. Miners' Assn Cornwall & Devon* (1873), pp.69–94 + map.

BARROW, G. & THOMAS, H. H. 1908. On the occurrence of metamorphic minerals in calcareous rocks in the Bodmin and Camelford areas, Cornwall. *Mineral. Mag.* **15**:113–123

—— & —— 1909 Some additional localities for idocrase in Cornwall. *Mineral. Mag.* **15**:238–240

BARTON, D. B. 1960 *The Redruth & Chasewater Railway 1824–1915*. Barton, Truro. (min. railway serving the Gwennap mines)

—— 1961 *A History of Copper Mining in Cornwall and Devon*. Barton, Truro. (1st ed.; 2nd ed., 1968).

—— 1963 *A Guide to the Mines of West Cornwall*. Barton, Truro.

—— 1964 *A Historical Survey of the Mines and Mineral Railways of East Cornwall and West Devon*. Barton, Truro.

—— 1965 *The Cornish Beam Engine*. Truro 1965 (new. ed. 1966)

—— 1967 *A History of Tin Mining and Smelting in Cornwall*. Barton, Truro.

—— (ed.) 1967 *Historic Cornish Mining Scenes Underground*. Truro 1967 (old photos by J. C. Burrow and H. W. Hughes)

—— 1968 *Essays in Cornish mining history*. vol.1. Truro. (The Cornish Miner in Fact and Fancy; Cornishmen and Australian Copper; Mine Names in the West of England; New Trumpet and Lovell United – the Anatomy of a Mine; The Techniques of Tin Smelting and Blowing; Water Engines in Cornish Mining)

—— 1970 *The Story of Cornwall's Engine-Houses*. Truro.

—— 1971 *Essays in Cornish mining history*. vol.2. Truro. (Some Characters in Cornish Mining; Some Cornish Blowing- and Melting-Houses; Newham, Calenick and Treyew – the Genesis of Reverberatory Tin Smelting; Arsenic Production in West Cornwall; Portreath and its Tramroad; Restronguet Creek Tin Works, 1871–1879)

BARTON, R. M. 1964 *An introduction to the geology of Cornwall*. Barton, Truro.

—— 1966 *A history of the Cornish china-clay industry*. Barton, Truro.

—— (ed.) 1970 / 1974 *Life in Cornwall in the early nineteenth century*. (1810–1835); *Life in Cornwall in the mid-nineteenth century*. (1835–1855); *Life in Cornwall in the late nineteenth century*. (1855–1875); *Life in Cornwall at the end of the nineteenth century*. (1876–1899). Barton, Truro (1970, 1971, 1972, 1974). (extracts from the *West Briton* newspaper, many on mines and mining).

BAWDEN, E. H. 1929 Killifreth Mine, Cornwall. *Mining Mag.* **1**:279–286

BAWDEN, M. G. 1962 The boron content of some Cornish rocks. *Proc. Ussher Soc.* **1**:11–13

BECHE, H. T. de la 1842 On the connection between geology and agriculture in Cornwall, Devon and West Somerset. *J. R. Agricultural Soc.* **3**:21–36

BECHER, J. J. 1682 *Alphabetum Minerale*, seu viginti quatuor theses chymicae de mineralium . . . . Truro. (unexpectedly little of interest, beyond the dedication. Another ed. (?), Frankfurt-am-Main 1689)

BEER, G. de 1966 Iktin – the tin port. *The Listener*, 1 Sept. 1966, pp.318–320.

BEER, K. E. 1978 Mineral deposits in the Variscides. pp.290–301, part of United Kingdom section (pp.263–317) In: *Mineral Deposits of Europe*, vol.1: Northwest Europe (eds. S. H. U. Bowie *et al.*). Inst. Min. Metall. and Mineral. Soc., London.

—— 1979 Paragenesis in the Variscan metallogenic province of Cornwall and Devon. In: *Freiberger Forschungshefte* C345 (Probleme der Paragenese), Topical Report of the Paragenetic Commission of IAGOD, Salamanca, Spain, April 13–16 1977, pp.15–27

—— & SCRIVENER, R. C. 1982 Metalliferous mineralisation. In: Durrance & Laming (eds.) 1982:117–145.

BENNETTS, S. 1887 The mining district of St Agnes. *Trans. Miners' Assn. & Inst. Cornwall* **1**(1):18–29.

BENNEY, D. E. 1972 *An Introduction to Cornish Watermills*. Barton, Truro.

BERGER, J. F. 1811 Observations on the Physical Structure of

Devonshire and Cornwall. *Trans. Geol. Soc.* 1:93–184

BERMAN, M. 1978 *Social Change and Scientific Organization / The Royal Institution 1799–1844*. London.

BIRD, R. H. 1974 *Britain's Old Metal Mines / A pictorial survey*. Barton, Truro.

—— 1977 *Yesterday's Golcondas / notable British metal mines*. Moorland Publishing Company, Buxton.

BLACKWELL, H. C. 1986 *From a Dark Stream*. / Being an account of how the history and geological formation of Cornwall affected its inhabitants and led to the great impact they had on many parts of the world [etc.]. Dyllansow Truran, Redruth. (well illustrated).

BLUETT, A. 1898 The Great Dolcoath. *Cornish Mag.* 1:168–181

—— 1899 Miners' Superstitions. *Cornish Mag.* 2:267–274

BOASE, G. C. 1890 *Collectanea Cornubiensia*: a collection of biographical and topographical notes relating to the county of Cornwall. Netherton & Worth, Truro. (limited ed., 130 copies; mineralogy, mines, mining, cols.1419–1423).

—— & COURTNEY, W. P. 1872–1882 *Bibliotheca Cornubiensis*. London, 3 vols: A-O, 1872; P-Z, 1878; Suppl. 1882

BOASE, H. S. 1822 On the Tin-ore of Botallack and Levant. *TRGSC* **2**:383–403

—— 1832 Contributions towards a knowledge of the geology of Cornwall. *TRGSC* **4**:166–474

—— *see* Gilbert 1838

BODE, G. (ed.) 1983 Mineralien Erzählen: Botallackit. *Magma* no.1, pp.14–17 (good photos)

BOOKER, F. 1967 *The Industrial Archaeology of the Tamar Valley*. David & Charles, Newton Abbot.

—— 1983 *Morwellham Quay in the Tamar Valley*. Morwellham Recreation Co., Tavistock.

BOOTH, B. 1968 Petrogenetic significance of alkali feldspar megacrysts and their inclusions in Cornubian granites. *Nature* **217**:1036–1038

BORLASE, G. S. 1832 Notice of some Records having reference to the commencement of Copper Mining in Cornwall and Devon. *TRGSC* **4**:486–489

BORLASE, Rev. W. 1749 Spar and Sparry Productions called Cornish Diamonds. *Phil. Trans. R. Soc. London* **46**:250

— 1758 *The Natural History of Cornwall*. Oxford. (also facsim. ed., London 1970, with Introduction by F. A. Turk, Biographical Note by P. A. S. Pool, and Appendix of additions from ms annotations by the author. These latter were published earlier in four parts, *J. R. Inst. Cornwall* 1864–1866).

—— 1767 extracts from letters to E. M. da Costa, concerning native tin. *Phil. Trans. R. Soc. London* **56**:35–39, 305–306

—— biography, *see* Pool, P. A. S.

BORLASE, W. C. 1874 *Historical sketch of the tin trade in Cornwall*, from the earliest period to the present day. William Brendon & Son, Plymouth.

BOTT, M. H. P. & SCOTT, P. 1964 Recent geophysical studies in southwest England. In: Hosking & Shrimpton (eds.) 1964, pp.25–44

—, HOLDER, A. P., LONG, R. E. & LUCAS, A. L. 1970 Crustal structure beneath the granites of southwest England. In: Newall, G. & Rast, N. (eds.) *Mechanism of igneous intrusion*. Geol. J., Spec. issue 2, pp.93–102

BOURNON, J. L. de 1801 Description of the Arseniates of Copper, and of Iron, from the County of Cornwall. *Phil. Trans. R. Soc. London* **91**:169–191 + figs. (see Chenevix 1801)

—— 1804 Description of a triple sulphuret of Lead, Antimony, and Copper, from Cornwall. *Phil. Trans. R. Soc. London* **94**:30–62 + figs. (bournonite)

—— 1813 *Catalogue de la Collection Minéralogique du Comte de Bournon* [etc.] avec un volume de planches. London.

— 1817 *Catalogue de la Collection Minéralogique particulière du Roi* [etc.]. Paris. (the catalogue proper, and the plates, are identical with those in Bournon 1813; but the 'Discours Préliminaire' (pp.i-cxiv) has been replaced by a Preface (pp.i-xv), and the two 'Observations' (pp.469–548) are omitted.)

BOWMAN, H. L. 1900 On Monazite and associated minerals from Tintagel, Cornwall. *Mineral. Mag.* **12**:358–362

—— 1911 On the occurrence of bertrandite at the Cheesewring Quarry, near Liskeard, Cornwall. *Mineral. Mag.* **16**:47–50

BRAITHWAITE, R. S. W. 1981 Turquoise crystals from Britain and a review of related species. *Mineral. Record* **12**:349–353

—— 1982 Wroewolfeite in Britain. *Mineral. Record* **13**:167–170,174 (Devon Friendship, p.168; IR spectra of related mins, some Cornish, p.170)

—— & COOPER, B. V. 1982 Childrenite in South-West England. *Mineral. Mag.* **46**:119–126

BRAMMALL, A. 1926 Gold and silver in the Dartmoor granite. *Mineral. Mag.* **21**:14–20

—— 1926 The Dartmoor granite. *Proc. Geol. Assn* **37**:26–38

—— 1928 Notes on fissure-phenomena and lode-trend in the Dartmoor Granite. *TRGSC* **16**:15–27

—— 1928 1928 Dartmoor detritals: a study in provenance. *Proc. Geol. Assn* **39**:27–48

—— & HARWOOD, H. F. 1923/1925 [minerals of the Dartmoor granite, various]. *Mineral. Mag.* **20**:20–26; 27–31; 39–53; 201–211; 319–330

—— & —— 1932 The Dartmoor granites: their genetic relationships. *Q. J. Geol. Soc. London* **88**:171–237

[BRANDE, W. T.] 1816 *A Descriptive Catalogue of the British Specimens deposited in the Geological Collections of the Royal Institution*. London.

BRAY, C. J. & SPOONER, E. T. C. 1983 Sheeted vein Sn-W mineralization and greisenization associated with economic kaolinization, Goonbarrow china clay pit, St Austell, Cornwall, England: geological relationships and geochronology. *Econ. Geol.* **78**:1064–1089

BRAY, [A. E.] Mrs. 1838 *Traditions, Legends, Superstitions, and Sketches of Devonshire on the Borders of the Tamar and the Tavy*, . . . in a series of letters to Robert Southey, Esq. 3 vols, 2nd ed., London 1838 (esp. vol.3:253–260, Letter XXXIX, quoting Edmund Pearse on the Virtuous Lady mine) (1st ed., diff. title but same pag., 1836; revised ed., 2 vols., 1879; Mrs Bray, née Stothard)

BRISTOW, C. M. 1977 A review of the evidence for the origin of the kaolin deposits in SW England. Proc. 8th International Kaolin Symposium and Meeting on Alunite, Madrid-Rome, Sept. 7–16, 1977.

—— 1979 *Geology of china clay in Cornwall*. Wheal Martyn Museum Booklet No. 2. St Austell China Clay Museum Ltd.

BRITTON, J. & BRAYLEY, E. W. 1831 *Cornwall Illustrated, in a Series of Views*, of Castles, Seats of the Nobility, Mines, [etc.]. From Original Drawings by Thomas Allom, &c. London. (ppb. reprint Barton, Truro 1968).

BROMLEY, A. V. 1975 Tin mineralization of Western Europe : is it related to crustal subduction?. *Trans. Inst. Min. Metall.* (Sect.B) **84**:B28–B30

—— 1976 A new interpretation of the Lizard Complex, south Cornwall, in the light of the ocean crust model. *J. Geol. Soc. London* **132**:114 (brief account of lecture)

—— 1976 Granites in mobile belts – the tectonic setting of the Cornubian batholith. *J. Camborne Sch. Mines* **76**:40–47

BROOKE, J. 1976 Wheal Buller. *J. Trevithick Soc.* no.4, pp.65–72

—— 1980 *Stannary Tales* / the Shady Side of Mining. Truro.

—— 1982 The last years of Devon Great Consols. *J. Trevithick Soc.* no.9, pp.69–72

BROOKS, M., MECHIE, J. & LLEWELLYN, D. J. 1983 Geophysical investigations in the Variscides of Southwest Britain. In: Hancock (ed.), pp.186–197

BROUGHTON, D. G. 1967 Tin workings in the eastern district of the parish of Chagford, Devon. *Proc. Geol. Assn* **78**:447–462

—— 1970 The Birch Tor and Vitifer tin mining complex. *Trans. Cornish Inst. Eng.* **24**:25–49 (+ discussion, pp.50–53)

BROWN, K. 1984 The incredible Ookiep copper mine: its riches, its railway and its Cornish engines. *J. Trevithick Soc.* no.11, pp.41–59

BROWN, L. C. 1964 Hemerdon Mine 1953–1954. *Trans. Cornish Inst. Eng.* **19**:26–36

BRYANT, N. 1871 On the Perran Iron Lode. *Rep. R. Cornwall Poly. Soc.* **38**:98–100

BUCKLEY, J. A. [1982] *A history of South Crofty mine.* Dyllansow Truran, Redruth (n.d.).

BUDGE, E. 1846 On the granitic and other associated rocks of Cornwall and Devon. *TRGSC* **6**:288–293

BURCHARD, U. & BODE, R. 1986 *Mineral Museums of Europe.* Walnut Hill Publishing Co. (esp. photos on pp. 111, 123, 126, 130, 131, 134, 135)

[BURR, F.] 1835 Descriptive notice of the Consolidated and United Mines. *Mining Rev.* **3**:17–61 (author's initials, F.B., only).

—— 1836 On the occurrence of the precious metals in Great Britain. *Mining Rev.* **3**:288–298

BURROW, J. C. & THOMAS, W. 1893 *'Mongst Mines and Miners.* Camborne. (repr. Truro 1965) (classic early underground photos.)

—— 1895 Photography in Mines. *TRGSC* **11**:621–633

—— see Thomas, H. 1896, and Barton (ed.) 1967

BURT, R. (ed.) 1969 *Cornish Mining / Essays on the Organisation of Cornish Mines and the Cornish Mining Economy.* David & Charles, Newton Abbot. (essays by J. Taylor 1814, 1837; C. Lemon 1838; J. Carne 1839; J. Sims 1849; L. L. Price 1891)

—— (ed.) 1972 *Cornwall's Mines and Miners.* Barton, Truro. (essays by G. Henwood, originally publ. in the *Mining J.* between 1857 and 1859)

—— 1977 *John Taylor / mining entrepreneur and engineer 1779–1863.* Moorlands Publishing Company, Buxton.

—— 1981 The Mineral Statistics of the United Kingdom: An Analysis of the Accuracy of the Copper and Tin Returns for Cornwall and Devon. *J. Trevithick Soc.* no.8, pp.31–46

—— 1984 *The British lead mining industry.* Dyllansow Truran, Redruth. (general, but much on Cornwall and Devon).

— , WAITE, P., & BROMLEY, R. 1984 *Devon and Somerset Mines /* Metalliferous and Associated Minerals 1845–1913. Exeter. (production statistics)

—— , —— , & BURNLEY, R. 1987 *Cornish Mines, Metalliferous and Associated Minerals. 1845–1913 (Mineral Statistics).* 562pp. University of Exeter in association with the Northern Mine Research Society.

—— & WILKIE, I. 1984 Manganese mining in the South-West of England. *J. Trevithick Soc.* no.11, pp.18–40

BURTON, A. 1976 *Josiah Wedgwood / A Biography.* Deutsch, London.

BUTLER, F. H. 1887 Note on Francolite. *Mineral. Mag.* **7**:164

—— 1907 On the occurrence of silver ore in the Perran mine, Perran Uthnoe, Cornwall. *Mineral. Mag.* **14**:385–388

—— , *see* Kinch 1886 (also, obit. in *Mineral. Mag.* **24**:279)

BUTT, P. S. 1974 (set of wall-chart illustrations) Major tin and copper mines in Cornwall 18th-20th Century; Railways of Cornwall past and present; Famous mining districts in Cornwall; General arrangement of a tin mine showing surface and underground workings, late 19th – early 20th Century (published by the author; sold at Truro Museum).

—— 1978 (ditto) Carn Brea Hill (ditto).

CALVERT, J. F. 1853 *The Gold Rocks of Great Britain and Ireland.* London. (ch.4, pp.74–94)

—— (biography) The late John Calvert / by one who knew him. *Mining J.* 1897, **67**:1332–1333. (1811/?14–1897; inconsistencies and exaggerations suggest it is from his own notes!)

CAMERON, J. 1951 The geology of Hemerdon wolfram mine, Devon. *Trans. Inst. Min. Metall.* **61**:1–14

CAMM, G. S. & HOSKING, K. F. G. 1984 The stanniferous placers of Cornwall, Southwest England. *Bull. Geol. Soc. Malaysia* **17**:323–356

—— & —— 1985 Stanniferous placer development on an evolving landsurface with special reference to placers near St Austell, Cornwall. *J. Geol. Soc. London* **142**:803–813

—— , TAYLOR, I., HARTWELL, P. & SCARBOROUGH, B. 1981 Carnon Valley, Cornwall, a placer tin deposit. *British Geologist* **7**(3):65–71

CAMPBELL SMITH, W. 1913 The mineral collection of Thomas Pennant (1726–1798). *Mineral. Mag.* **16**:331–342. (Borlase, p.336; Pennant's collection and ms catalogue are in the BM(NH)).

—— 1969 A history of the first hundred years of the mineral collection in the British Museum / with particular reference to the work of Charles Konig. *Bull. Brit. Mus. (Nat. Hist.)*, Hist. Ser. **3**(8):237–259

—— 1978 Early mineralogy in Great Britain and Ireland. *Bull. Brit. Mus. (Nat. Hist.)*, Hist. Ser. **6**(3):49–74

CANN, F. C. 1917 The mines, lodes and minerals of the Stennack Valley, St Ives. *Trans. Cornish Inst. Min. Mech. Metall. Eng.* **5**:11–25.

CAREW, R. 1602 *Survey of Cornwall.* (reprinted 1723; London 1769; 1811, with notes by T. Tonkin (ed. de Dunstanville); with other works, edited by Halliday, London 1953) (index in Gilbert 1838)

CARNE, J. 1818 On the Discovery of Silver in the Mines of Cornwall. *TRGSC* **1**:118–126

—— 1822 On the relative Age of the Veins of Cornwall. *TRGSC* **2**:49–128

—— 1822 On the Mineral Productions and the Geology of the Parish of St. Just. *TRGSC* **2**:290–358 + map (v. full account)

—— 1827 On the Period of the Commencement of Copper Mining in Cornwall; and, On the Improvements which have been made in Mining. *TRGSC* **3**:35–85

—— 1827 On the granite of the western part of Cornwall. *TRGSC* **3**:208–246

—— 1829 A Description of the Stream-Work at Drift Moor, near Penzance. *TRGSC* **4**:47–56

—— 1832 An account of the Discovery of some varieties of Tin-Ore in a Vein, which have been considered peculiar to Streams; with remarks on Diluvial Tin in general. *TRGSC* **4**:95–112. (wood tin; Garth mine, pp.97–100)

—— 1839 Statistics of the tin mines in Cornwall, and of the consumption of tin in Great Britain. *J. Statist. Soc.* **2**:260–268 (reprinted in Burt 1969)

—— 1846 On the pseudomorphous minerals found in Cornwall, illustrative of a replacement of one mineral substance by another. *TRGSC* **6**:24–31

CARSWELL, J. 1950 *The Prospector / Being the Life and Times of Rudolf Erich Raspe (1737–1794).* Cresset Press, London.

CHALLINOR, J. 1971 *The History of British Geology / A Bibliographical Study.* David & Charles, Newton Abbot.

CHALMERS-HUNT, J. M. 1976 *Natural History Auctions 1700–1972* (A Register of Sales in the British Isles). Sotheby, London.

CHANDLER, C. & ISAAC, K. P. 1982 The geological setting, geochemistry and significance of Lower Carboniferous basic volcanic rocks in Central South-west England. *Proc. Ussher Soc.* **5**:279–288

CHAROY, B. 1979 Greisenisation, mineralisation et fluides associes Cligga Head, Cornwall (sud-ouest de l'Angleterre). *Bull. Mineral.* **102**:633–641

—— 1981 Post-magmatic processes in south-west England

(org. chairman) 1985.

HOWIE, R. A. 1965 Bustamite, rhodonite, spessartine, and tephroite from Meldon, Okehampton, Devon. *Mineral. Mag.* **34**:249–255

—— 1975 Tourmaline. *Proc. Ussher Soc.* **3**:209–213

—— *see* Chaudhry, M. N., also Moore, F.

HUDSON, K. [1969] *The History of English China Clays / Fifty Years of Pioneering and Growth.* David & Charles, Newton Abbot (n.d.).

HUNT, J. D. 1985 'Curiosities to Adorn Cabinets and Gardens'. ch.23 In: Impey & Macgregor (eds.) 1985:193–203

HUNT, R. & PHILLIPS, J. 1842 On the electricity of mineral veins. *Rep. R. Cornwall Poly. Soc.* **9**:157–164

—— & —— 1843 Experiments and observations on the electricity of mineral veins. *Rep. R. Cornwall Poly. Soc.* **10**:26–27

—— 1868 The Economic Geology of Devonshire and Cornwall in 1868. *J. Bath & West of England Soc.* [etc.] **16**:67–90

—— 1884 *British Mining* / A Treatise on the History, Discovery, Practical Development and Future Prospects of Metalliferous Mines in the United Kingdom. London 1884. (2nd ed. London 1887)

—— biography, *see* Pearson 1976; for mining statistics, *see* Geol. Surv. section (below)

HUTCHINSON, A. 1900 On Stokesite, a new mineral containing tin, from Cornwall. *Mineral. Mag.* **12**:274–281

—— 1903 The chemical composition and optical characters of chalybite from Cornwall. *Mineral. Mag.* **13**:209–216

HUTTON, D. H. W. & SANDERSON, D. J. (eds.) 1984 *Variscan tectonics of the North Atlantic region.* Geol. Soc. London.

IMPEY, O. & MACGREGOR, A. (eds.) 1985 *The Origins of Museums / The Cabinet of Curiosities in Sixteenth- and Seventeenth-Century Europe.* Oxford.

ISAAC, K. P., TURNER, P. J., & STEWART, I. J. 1982 The evolution of the Hercynides of central SW England. *J. Geol. Soc. London* **139**:521–531

JACKSON, N. J. & ALDERTON, D. H. M. 1974 Discordant calc-silicate bodies in the Botallack area. *Proc. Ussher Soc.* **3**(1):123–128

—— 1974 Gryll's Bunny, a 'tin floor' at Botallack. *Proc. Ussher Soc.* **3**:186–188.

—— 1975 'Carbonas' – a review. *Proc. Ussher Soc.* **3**:218–219 (also The Levant Mine carbona, **3**:220–225, 427–429)

—— & RANKIN, A. H. 1976 Fluid inclusion studies at St. Michael's Mount. *Proc. Ussher Soc.* **3**:430–434

—— , MOORE, J. McM., & RANKIN, A. H. 1977 Fluid inclusions and mineralisation at Cligga Head. *J. Geol. Soc. Lond.* **134**:343–349

—— 1978 The Halvosso pegmatites. *Proc. Ussher Soc.* **4**:190–191

—— , HALLIDAY, A. N., SHEPPARD, S. M. F. & MITCHELL, J. G. 1982 Hydrothermal activity in the St. Just mining district, Cornwall, England. In: Evans (ed.) 1982:137–179

JAMES, C. C. 1944 Mine temperatures in West Cornwall. *TRGSC* **17**:164–173

—— [1945] Great Wheal Vor. *TRGSC* **17**:194–207

—— [1945] Uranium Ores in Cornish Mines. *TRGSC* **17**:256–268

—— [1949] *A History of the Parish of Gwennap in Cornwall.* Penzance [n.d., ?1949] (contains much information on this famous mining parish)

JARS, G. 1781 *Voyages métallurgiques*, ou recherches et observations sur les mines de cuivre . . . les mines de l'étain . . . faites en 1758, 1765, jusques & compris 1769, en Allemagne, en Suede, Angleterre, Norvege, Tirol, Liege, & en Hollande. 3 vols, Paris 1781. (vol.3, p.86,186–206,222–223,522–537)

JENKIN, A. K. H. 1925 Cornish Mining Yesterday, Today, and Tomorrow. *Mining Mag.* **33**:376–379. (summary of a lecture).

—— 1927 / 1948 *The Cornish Miner.* 2nd ed., Allen & Unwin, London. (1st ed. 1927; 1972 reprint by David & Charles, Newton Abbot).

—— 1930 The History of Great Wheal Vor. *J. R. Inst. Cornwall* **23**:300–324

—— 1951 *News from Cornwall* / With a Memoir of William Jenkin. Westaway Books, London. (letters, 1790–1820).

—— 1959 The Rise and Fall of Wheal Alfred. *J. R. Inst. Cornwall* (n.s.) **3**:124–137.

—— 1961–1970 *Mines and Miners of Cornwall*:
I. Around St Ives (1961)
II. St Agnes – Perranporth (1962)
III. Around Redruth (1962)
IV. Penzance – Mount's Bay (1962)
V. Hayle, Gwinear, & Gwithian (1963)
VI. Around Gwennap (1963)
VII. Perranporth – Newquay (1963)
VIII. Truro to the Clay District (1964)
IX. Padstow, St Columb & Bodmin (1964)
X. Camborne – Illogan (1965)
XI. Marazion, St Hilary & Breage (1965)
XII. Around Liskeard (1966)
XIII. The Lizard – Falmouth – Mevagissey (1967)
XIV. St Austell to Saltash (1967)
XV. Calstock, Callington & Launceston (1969)
XVI. Wadebridge, Camelford & Bude (1970)

—— & ORDISH, G. Index to *Mines and Miners of Cornwall* (n.d) (separately paginated; I–XIV publ. by Barton, Truro; XV, XVI, and Index by the Federation of Old Cornwall Societies).

—— 1974 *Mines of Devon* / Vol.1 The Southern Area. David & Charles, Newton Abbot.

—— 1978 *Wendron Tin.* Wendron Forge Ltd., Helston.

—— 1981 *Mines of Devon* / North and East of Dartmoor. Devon Library Services, Exeter.

JOHNS, R. K. 1986 *Cornish Mining Heritage.* South Australia Dept. of Mines and Energy, Special Publ. no.6.

JONES, B. 1974 *Follies & Grottoes.* 2nd ed., London. (repr. 1979; 1st ed. 1953)

JORDANOVA, L. J. & PORTER, R. S. (eds.) 1979 *Images of the Earth* / Essays in the History of the Environmental Sciences. Brit. Soc. Hist. Sci., Monograph 1.

KEAST, J. 1982 *'The King of Mid-Cornwall' / The life of Joseph Thomas Treffry (1782–1850).* Dyllansow Truran, Redruth. (much on mining, quarrying, and transport).

KINCH, E. & BUTLER, F. H. 1886 On a New Variety of Mineral from Cornwall; with a Note . . . by H. A. Miers. *Mineral. Mag.* **7**:65–70 (dufrenite).

—— 1888 On Dufrenite from Cornwall. *Mineral. Mag.* **8**:112–115

KEAR, D. 1952 Mineralization at Castle an Dinas Wolfram Mine, Cornwall (discussion). *Trans. Inst. Min. Metall.* **61**:408–410

KETTANEH, Y. A. & BADHAM, J. P. N. 1978 Mineralization and paragenesis at the Mount Wellington Mine, Cornwall. *Econ. Geol.* **73**:486–495

KINGDON, [A. S.] 1867 The Silver Mines at Combmartin. *Rep. Trans. Devon. Assn* **2**:190–199

KINGSBURY, A. W. G. 1954 New occurrences of rare copper and other minerals [in Devon and Cornwall]. *TRGSC* **18** (part 4 for 1952), pp.386–406

—— 1956 The rediscovery of churchite in Cornwall. *Mineral. Mag.* **31**:282

—— 1957 Rockbridgeite from Cornwall and Devon. *Mineral. Mag.* **31**:429; A new occurrence of nadorite in Cornwall. **31**:499

—— 1958 Two beryllium minerals new to Britain: euclase and herderite. *Mineral. Mag.* **31**:815–817

—— 1959 A new occurrence of fluellite in Cornwall, and its paragenesis. *TRGSC* **19**:42–51 (Goonvean claypit)

—— 1961 Beryllium minerals in Cornwall and Devon: helvine, genthelvite, and danalite. *Mineral. Mag.* **32**:921–940

—— 1964 Some minerals of special interest in South-West England. In: Hosking & Shrimpton (eds.) 1964:243–266.

—— 1965 Tellurbismuth and meneghinite, two minerals new to Britain. *Mineral. Mag.* **35**:424–426

—— 1966 Sir Arthur Russell [memorial] *Mineral. Mag.* **35**:673–677; Memorial of Sir Arthur Russell. *Amer. Mineral.* **53**:596–599

—— *see* also under Davis 1965, & Macfadyen 1970; for memorial, see Embrey 1973

KINGSTON, J. T. 1828 Account of the iron mine at Haytor. *Phil. Mag.* (n.s.) **3**:359–365.

KIRBY, G. 1978 Layered gabbros in the eastern Lizard, Cornwall, and their significance. *Geol. Mag.* **115**:199–204.

KLAPROTH, M. H. (and Groschke, J. G., trans.) 1787 *Observations relative to the Mineralogical and Chemical History of the Fossils of Cornwall.* London. (transl., with introductory pages, of 'Mineralogisch-chemischer Beytrag zur Naturgeschichte Cornwallischer Mineralien' von Klaproth. *Schriften der Gesellschaft naturforschender Freunde zu Berlin* 1787, **7**:141–196 + Taf.2)

— 1808 Zerlegung des Wavelits. A. Wavelit von Barnstapel. *Magazin. Gesellschaft naturforschender Freunde zu Berlin* 1808, p.4 (name Devonite given by Wm Thompson to [wavellite] specimens sent to Klaproth)

KNEEBONE, D. 1983 *'Fish, Tin, and Copper'.* Dyllansow Truran, Redruth.

LAKE, W. (publ.) *Parochial History* [etc.], 1867–1872 *see* Polsue.

LAW, R. J. 1976 A glimpse of the Cornish mineral industry in 1873. *J. Trevithick Soc.* no.4, pp.57–62

LAWRENCE, A. 1898 Romance in Hard Metal / an interview with Mr George Tangye. *Cornish Mag.* **1**:339–356

LEAN, T. & BROTHER [Joel Lean] 1839 *Historical Statement of the Improvements made in the Duty performed by the Steam Engines in Cornwall,* from the commencement of the publication of the Monthly Reports. Simpkin, Marshall & Co., London.

LEE, G. S. 1968 Prospecting for tin in the sands of St Ives Bay, Cornwall. *Trans. Inst. Min. Metall.* (Sect.A) **77**:A49–A64

LEECH, J. G. C. 1929 St. Austell detritals. *Proc. Geol. Assn* **40**:139–146

LEES, P. B. 1914 On the geological history of William's Lode in King Edward Mine, Camborne. *TRGSC* **13**:611–641

LEESE, C. E. & SETCHELL, J. [1938] Notes on Delabole Slate Quarry. *TRGSC* **17**:41–48

[LEIFCHILD, J. R.] 1855 *Cornwall: its Mines and Miners.* / With sketches of scenery, designed as a popular introduction to metallic mines. Longman *et al.*, London. (2nd edn, 1857: repr. 1968, Cass & Co., London).

LETTSOM, W. G. *see* Greg & ——, 1858; also Nevill 1872

LEVERIDGE, B. E., HOLDER, M. T. & DAY, G. A. 1984 Thrust nappe tectonics in the Devonian of south Cornwall and the western English Channel. In: Hutton & Sanderson (eds) 1984:103–112

LEVINE, J. 1977 *Dr Woodward's Shield:* [etc.]. Berkeley.

LÉVY A. 1824 On a new Mineral Substance. *Ann. Phil.* (n.s.) **8**:241–243. (more than one. Fluellite pp.242–243 + Pl.XXXII, fig.8).

—— 1827 On the origin of the crystalline forms of haytorite. *Phil. Mag.* (n.s.) **1**:43–46.

—— 1838 *Description d'une Collection de Minéraux, formée par M. Henri Heuland, et appartenant a M. Ch. Hampden Turner,* [etc.]. 3 vols. + atlas of plates. London. (intro. by Heuland)

LEWIS, G. R. 1908 *The Stannaries / A study of the medieval tin miners of Cornwall and Devon.* (reprint, Barton, Truro 1965)

—— *see also* Page (ed.), 1906

LEWIS, M. J. T. 1960 *The Pentewan Railway 1829–1918.* Barton, Truro.

LEWIS, W. J. 1882 On a crystal of Stephanite from Wheal Newton. *Proc. Camb. Phil. Soc.* **4**(4):240–245 (also Z. *Kryst.* **7**:575)

—— & HALL, A. L. 1900 On some remarkable Composite Crystals of Copper Pyrites from Cornwall. *Mineral. Mag.* **12**:324–332

LEWIS, W. S. 1923 *West of England Tin Mining.* Exeter.

LINTON, D. L. 1955 The problem of tors. *Geogr. J.* **121**:470–487

LISTER, C. J. 1978 Luxullianite *in situ* within the St. Austell granite, Cornwall. *Mineral. Mag.* **42**:295–297 (also, **43**:442–443)

—— 1984 Xenolith assimilation in the granites of south-west England. *Proc. Ussher Soc.* **6**:46–53

LLEWELLYN, B. 1946 A survey of the deeper tin zones in a part of the Carn Brea Area, Cornwall. *Trans. Inst. Min. Metall.* **55**:505–557

LLOYD, D. & BARSTOW, R. W. 1983 A new occurrence of aurichalcite and hemimorphite from Cornwall. *J. Russell Soc.* **1**(2):23–24.

LLOYD, G. E. & WHALLEY, J. S. 1986 The modification of chevron folds by simple shear: examples from north Cornwall and Devon. *J. Geol. Soc. London* **143**:89–94

LONG, L. E. 1962 Some isotopic ages from south-west England. In: Coe, K. (ed.) *Some aspects of the Variscan Fold Belt.* Manchester Univ. Press 1962:129–134

LUDLAM, H., *see* Rudler 1905, also Nevill 1872; obit. in *Mineral. Mag.* **4**:132

LUGARO, G. 1926 Sulla Bismutinite di St. Agnes (Cornovaglia). *Rendiconti R. Accad. naz. Lincei* (ser.6) **3**:416–419.

LYSONS, Revd D. & S. [1814] *Magna Britannia* (vol. III). Topographical and Historical Account of the County of Cornwall. London (n.d.). (completion date, Aug. 1814, noted in *Gentleman's Mag.* **84**(2):136). (section 'Natural History, Fossils and Minerals', pp.cxciv-cxcviii, by 'a friend' i.e., J. Hawkins)

—— 1822 *Magna Britannia: being a concise topographical account of the several Counties of Great Britain. Volume the Sixth, containing Devonshire.* London. (section 'Natural History, Minerals', pp.cclxv-cclxx, by J. Hawkins).

—— *see* Steer 1966 (correspondence).

MacALISTER, D. A. 1904 A cross-section and some notes on the tin and copper deposits of Camborne, with special reference to the limits of productive ore ground. *TRGSC* **12**:773–795 + 5 figs.

—— 1909 Note on the association of cassiterite and specular iron in the lodes of Dartmoor. *Geol. Mag.* **6**:402–409

—— *see also* section on Geol. Survey publications

MacCULLOCH, J. 1814 On the granite tors of Cornwall. *Trans. Geol. Soc. London* **2**:66–78

MacFADYEN, W. A. 1970 (with contributions by A. W. G. Kingsbury) *Geological Highlights of the West Country.* A Nature Conservancy Handbook. London.

MacGREGOR, A. (ed.) 1983 *Tradescant's Rarities / Essays on the foundation of the Ashmolean Museum 1683 with a catalogue of the surviving early collections.* Oxford.

—— 1983 Collectors and Collections of Rarities in the Sixteenth and Seventeenth Centuries. In: Macgregor (ed.) 1983:77–97

—— *see* Impey, O.

MacGREGOR, W. (Mr) error for Gregor, Rev. W. (q.v.) in Greg & Lettsom (1858:379) and in *Dana's Syst. Min.* 5th ed. (1868:793), 6th ed. (1892:307)

MACK, M. 1969 *The Garden and the City: Retirement and Politics in the Late Poetry of Pope.* U. of Toronto Press. (account of Alexander Pope's mineral grotto at Twickenham, chs. 2 and 3, apps. A and C; details from Searle 1745)

MacLAREN, M. 1917 The geology of the East Pool Mine. *Mining Mag.* **16**:245–252

MacLEAN, J. 1874 The tin trade of Cornwall in the reign of Elizabeth and James compared with that of Edward I. *J. R. Inst. Cornwall.*, **4**:187–190

MAJENDIE, A. 1818 A Sketch of the Geology of the Lizard District. *TRGSC* **1**:32–37

—— 1818 Contributions towards a knowledge of the Geological History of Wood Tin. *TRGSC* **1**:237–239

MANNING, D. A. C. 1981 The application of experimental studies in determining the origin of topaz-quartz-tourmaline rock and tourmaline-quartz rock (Roche Rock). *Proc. Ussher Soc.* **5**:121–127

—— 1983 Disseminated tin sulphides in the St. Austell granite. *Proc. Ussher Soc.* **5**:411–416

—— & EXLEY, C. S. 1984 The origins of late-stage rocks in the St. Austell granite – A re-interpretation. *J. Geol. Soc. London* **141**:581–593

—— 1985 A comparison of the influence of magmatic water on the form of granite-hosted Sn-W deposits and associated tourmalinisation from Thailand and southwest England. In: Halls (org. chairman) 1985:203–212.

MARTIN, J. S. 1895 Micaceous iron ore near Bovey Tracey. *Trans. Manchester Geol. Soc.* **23**:161–163

MASKELYNE, [M. H.] N. S. & LANG, V. von 1863 Mineralogical Notes. 1. On Connellite. *Phil. Mag.* (ser.4) **25**:39–41.

—— 1865 On New Cornish Minerals of the Brochantite Group. *Proc. R. Soc. London* **14**:392–400 (langite, waringtonite)

—— 1871 On andrewsite. *Rep. Brit. Assn* (1871), p.75

—— & FLIGHT, W. 1871–1872 Mineralogical Notices. 3. Francolite, Cornwall. *J. Chem.* Soc. (ser.2) **9**:3–5; 5. Vivianite. **9**:6–9; 6. Cronstedtite, **9**:9–12. 11. Uranite, **10**:1054–1055.

—— 1875 On Andrewsite and Chalkosiderite. *J. Chem. Soc.* (ser.2) **13**:586–591.

——, *see* Field 1877

MATON, W. G. 1797 *Observations relative chiefly to the Natural History, Picturesque Scenery, and Antiquities, of the Western Counties of England, Made in the Years 1794 and 1796*. 2 vols. Salisbury. (earliest publ. geol. map of southwest England).

MAWE, J. 1818 On the Tourmalin and Apatite of Devonshire. *J. Sci. Arts, London* **4**:369–372

MAYNARD, J. 1874 Remarks on two cross-sections through Carn Brea Hill and the neighbouring mines. *Rep. Miners' Assn Cornwall & Devon* (1873), pp.43–54.

—— 1874 Mines of the Illogan District. *Rep. R. Cornwall Poly. Soc.* (1874), p.84

—— 1876 Note on a cross section from Cook's Kitchen to Wheal Emily Henrietta. *Rep. Miners' Assn Cornwall & Devon* (1875), pp.65–67

McCARTNEY, P. J. 1977 *Henry De la Beche: observations on an observer*. Cardiff.

McGREGOR, W. 1791 error for Gregor, W. (q.v.)

McLINTOCK, W. F. P. 1923 On the occurrence of petalite and pneumatolytic apatite in the Meldon aplite, Okehampton, Devonshire. *Mineral. Mag.* **20**:140–150

McPHERSON, G. & LAMB, T. 1921 Platinum-bearing Rocks in the Lizard District. *Geol. Mag.* **58**:512–514

MEADE, R. 1882 *The coal and iron industries of the United Kingdom*. Crosby Lockwood & Co., London. (Devonshire iron industries, pp.688–701; Cornwall iron industries, pp.702–718)

MERRETT, C. 1678 A Relation of the Tinn-Mines, and working of Tinn in the County of Cornwal. *Phil. Trans. R. Soc. London* **12**:949–952 (Mundick, Cornish Diamonds, Godolphin Ball)

MESSENGER, M. J. 1978 *Caradon and Looe: The Canal, Railways and Mines*. Twelveheads Press.

METCALFE, J. E. 1969 *British Mining Fields*. Inst. Min. Metall., London.

MICHELL, F. 1978 *Annals of an ancient Cornish town / being notes on the history of Redruth*. 2nd ed. (2nd impression 1985, ppb., cover title *Annals of an Ancient Cornish Town – Redruth*) Dyllansow Truran, Redruth. (1st ed. 1948). (fascinating detail, incl. much on mining, arranged in sections by years).

MICHELL, F. B. 1978 Ore dressing in Cornwall 1600–1900. *J. Trevithick Soc.* no.6, pp.25–52

—— 1980 The introduction of the plunger pole or force pump. *J. Trevithick Soc.* no.7, pp.34–36

MITCHELL, S., ARGALL, W. H. & EUDEY, J. 1871 Account of an excursion of the Breage, St Just, Redruth and Carharrack classes to the clay and tin works of Carclaze, St. Austell, on the 2nd of August 1870. *Rep. Miners' Assn Cornwall & Devon* (1870), pp.33–39

MIERS, H. A. 1884 The crystallography of bournonite. *Mineral. Mag.* **6**:59–79.

—— 1885 On monazite from Cornwall, and connellite. *Mineral. Mag.* **6**:164–170

—— 1889 Mineralogical notes. Polybasite; aikinite; quartz; cuprite; the locality of turnerite. *Mineral. Mag.* **8**:204–209

—— 1894 Spangolite. *Mineral.Mag.* **10**:273–277 (also *Nature* 1893, **48**:426–427)

—— 1897 On some British pseudomorphs. *Mineral. Mag.* **11**:263–285

—— & PRIOR, G. T. 1892 Danalite from Cornwall. *Mineral. Mag.* **10**:10–14

MILLER, J. A. & MOHR, P. A. 1964 Potassium-argon measurements on the granites and some associated rocks from SW England. *Geol. J.* **4**:105–126

MILLER, N. 1982 *Heavenly Caves / Reflections on the Garden Grotto*. London 1982 (general, no mention of Menabilly or Pendarves)

MILLER, W. A. 1864 Chemical examination of a hot spring containing caesium and lithium in Wheal Clifford, Cornwall. *Chem. News* **10**:181–182 (also *Mining & Smelting Mag.* **6**:197–198)

MITCHELL, A. H. G. 1974 Southwest England granites : magmatism and tin mineralization in a post-collision tectonic setting. *Trans. Inst. Min. Metall.* (Sect.B) **83**:B95–B97

MOISSENET, M. L. 1862 Etudes sur les filons du Cornouailles et du Devonshire. *Ann. des Mines* (ser.6) **3**:161–171.

—— 1874 *Etudes sur les filons du Cornwall*: Parties riches des filons, structure de ces parties et leur relation avec les directions des systemes stratigraphiques. Dunod, Paris. (11 pl., separately bound)

—— 1877 Observations on the rich parts of the lodes of Cornwall. (transl. by J.H. Collins). Lake & Lake, Truro.

MOORE, F. 1977 The occurrence of topaz-rich greisens at St. Michael's Mount, Cornwall. *Proc. Ussher Soc.* **4**:49–56

—— & HOWIE, R. A. 1979 Geochemistry of some Cornubian cassiterite. *Mineralium Deposita* **14**:103–107

MOORE, J. McM. 1975 A mechanical interpretation of the vein and dyke systems of the SW England orefield. *Mineralium Deposita* **10**:374–388

—— 1977 Exploration prospects for stockwork tungsten-tin ores in Southwest England. *Mining Mag.* **136**:97–103

—— & JACKSON, N. J. 1977 Structure and mineralization in the Cligga granite stock, SW England. *J. Geol. Soc. London* **133**:467–480

—— 1982 Mineral zonation near the granitic batholiths of south west and northern England, and some geothermal analogues. In: Evans (ed.) 1982:229–241

—— & CAMM, S. 1983 Interactive enhancement of Landsat imagery for structural mapping in tin-tungsten prospecting: A case history of the S.W. England orefield (U.K.). In: *Int. Symposium on Remote Sensing of Environment*, Second Thematic Conference, Fort Worth, Texas, 1982, pp.727–740

MOORE, (Rev Mr) 1755 Verses occasion'd by seeing the Fossilry at Penderves [sic] in Cornwall. Inscrib'd to Mrs Percival. *Gentleman's Mag.* **25**:567–568 (composed c.1747, acc. to C. S. Gilbert)

MORRISON, T. A. 1980 *Cornwall's Central Mines / The Northern District 1810–1895*. Penzance.

—— 1983 *Cornwall's Central Mines / The Southern District 1810– 1895*. Penzance.

MOUNT, M. 1985 Geevor mine: a review. In: Halls (org. chairman) 1985:221–238

MOYLE, M. P. 1822 On the temperature of the Cornish Mines. *TRGSC* **2**:404–415

MOYLE, S. 1840 On the ventilation of Mines. *Rep. R. Inst. Cornwall* 1840, pp.57–60

MURRAY (publ.) 1850 *A Hand-book for Travellers in Devon & Cornwall*. Murray, London. (with maps) (several other editions)

MURRAY, D. 1904 *Museums / Their History and their Uses* [etc.]. 3 vols. Glasgow. (fate of Roy. Inst. coll., vol.2:351).

NATIONAL TRUST 1984 *The National Trust Atlas*. George Philip & Son and National Trust (1st ed., 1981; 2nd ed. 1984, repr. 1986)

NEIL, J. S. [1908] *British Minerals and where to find them*. Murby, London (n.d.).

NEVILL, W. 1872 *Descriptive Catalogue of Minerals, being the collection of William Nevill, F.G.S., Godalming, Surrey*. London. (incorporated W. G. Lettsom's coll.; sold to H. Ludlam in 1877, now in Geol. Mus. – BM(NH))

NEWELL, E. 1986 Interpreting the Cornish copper standard. *J. Trevithick Soc.* no.13, pp.36–45

NOALL, C. 1970 *Levant: The Mine Beneath the Sea*. Barton, Truro.

—— 1972 *Botallack*. Barton, Truro.

—— 1973 *The St. Just Mining District*. Barton, Truro.

—— 1982 *The St. Ives mining district*. Vol.1. Dyllansow Truran, Redruth. (the author died before completing vol.2).

—— 1983 *Geevor*. Geevor Tin Mines Plc, Pendeen. Penzance. (ppb.).

NOBLE, G. 1872 Remarks on Mineral Veins in the Parish of Constantine. *Rep. Miners' Assn Cornwall & Devon* (1871), pp.45–46 (also *Rep. R. Cornwall Poly. Soc.* (1871), pp.74–76)

NOELTING, J. 1887 Ueber das Verhältniss der sogenannten Schalenblende zur regulären Blende und zum hexagonalen Würtzit. Inaug. Diss., Kiel. (Not seen; abstr. in *Neues Jahrb. Min. (Ref.)* 1888, **1**:205; *Zeits. Kryst.* **17**:220). (Wurtzite from Liskeard, Wh. Unity, and Tavistock)

NORDEN, J. 1728 *Speculi Britanniae Pars: A Topographical and Historical Description of Cornwall. With A Map of the County and each Hundred*; [etc.] London 1728. (John Norden (1548–1626); 1728 publ. little or no diff. from ms., completed c.1606) (ppb. reprint Graham 1966; not facsim., different pagination; the date '1650' on the cover seems to be a typo. for 1605).

ORDISH, H. G. 1967 / 1968 *Cornish Engine-Houses, a Pictorial Survey*; and *Cornish Engine-Houses, a Second Pictorial Survey*. Barton, Truro.

—— *see* Jenkin, A. K. H.

OWEN, T. R. 1976 *The Geological Evolution of the British Isles*. Pergamon Press, Oxford.

PAGE, W. (ed.) 1906 *The Victoria History of the County of Cornwall*. vol.1. London. (sections on: Geology, J. B. Hill; Granite Quarrying, China Clay, T. Taylor; Tin and Copper Mining, G. R. Lewis; Slate Quarrying, J. Hockaday; etc.)

PALMER, J. & NIELSON, R. A. 1962 The origin of tors on Dartmoor, Devonshire. *Proc. Yorks. Geol. Soc.* **33**:315–340

PALMER, M. & NEAVERSON, P. 1987 *The Basset Mines / Their History & Industrial Archaeology*. Northern Mines Research Soc., Sheffield. British Mining No. 32.

[PARIS, J. A.] 1816 *A Guide to the Mount's Bay and the Land's End*; . . . Penzance 1816. (pp.ix + 156. 2nd ed., much enlarged, pp.xix + 272, 'by a Physician', London 1824)

—— 1818 *A Memoir of the Life and Scientific Labours of the late Rev. William Gregor, A.M.* London.

—— 1818 Gregorite (Menacchanite) discovered at Lanarth. *TRGSC* **1**:226–227

PASCOE, W. A. 1945 The Foweymoor District. *Mining Mag.* **72**:210–212

PASCOE, W. H. [1982] *C.C.C. / The History of the Cornish Copper Company*. Dyllansow Truran, Redruth (n.d.; ppb.).

PATTISON, S. R. 1855 A day in the North Devon Mineral District. *TRGSC* **7**:223–227

PAYTON, P. J. 1978 *Pictorial History of Australia's Little Cornwall*. Rigby Limited, Adelaide.

—— 1984 *The Cornish Miner in Australia* / (Cousin Jack Down Under). Dyllansow Truran, Redruth.

PEARCE, R. 1861–1863 On some interesting minerals recently found in a few of the Cornish Mines. *Rep. R. Inst. Cornwall.* **43**:34–36; **44**:19–20

—— 1872 Notes on the occurrence of cobalt in connection with the tin ores of Cornwall. *J. R. Inst. Cornwall* **4**:81–83

—— 1878 Note on pitchblende in Cornwall. *TRGSC* **9**:103–104

PEARSON, A. 1976 *Robert Hunt, F.R.S. (1807–1887)*. Penzance.

PENALUNA, W. (publ.) 1848 *An Historical Survey of the County of Cornwall*; [etc.] 2 vols., 2nd ed., Penaluna, Helston 1848 (1st ed. 1838; Penaluna pubd. Hitchins & Drew's 'History', 1824)

PENBERTHY, J. 1846 Notes on the discovery of a quantity of pitchblende at Providence Mines, Near St. Ives. *TRGSC* **6**:106–107

PENDARVES, E. W. W. 1827 Notice of the Native Copper of Condorrow Mine. *TRGSC* **3**:333–334 (i.e. Condurrow mine)

PENDERILL-CHURCH, J. 1972 *William Cookworthy 1705–1780* / A study of the pioneer of true porcelain manufacture in England. Barton, Truro.

PENGELLY, W. 1856 On the Beekites found in the Red Conglomerates of Torbay. *TRGSC* **7**:309–315

PENHALE, J. 1962 *The Mine Under the Sea*. Lake, Falmouth. (originally published serially in the *Cornish Mag.* 1961/62; J. Penhale is *nom de plume* of R. Harry) (Levant mine)

PENHALLURICK, R. D. 1986 *Tin in Antiquity* / its mining and trade throughout the ancient world with particular reference to Cornwall. Institute of Metals, London.

PEPPER, J. H. 1861 *The Playbook of Metals*: including personal narratives of visits to coal, lead, copper, and tin mines; [etc.] London & New York.

PHEMISTER, J. 1940 Note on an occurrence of bertrandite and beryl at the South Crofty mine, Cornwall. *Mineral. Mag.* **25**:573–578

PHILLIPS, F. C. 1964 Metamorphism in South-west England. In: Hosking & Shrimpton (eds.) 1964:185–200

PHILLIPS, J. A. 1844. On the electricity of mineral veins. *Rep. R. Cornwall Poly. Soc.* **11**:54–59

—— 1871 On the connexion of certain phenomena with the origin of mineral veins. *Phil. Mag.* (ser.4) **42**:401–413.

—— 1873 On the composition and origin of the waters of a salt spring in Wheal Seton, with a chemical and microscopical examination of certain rocks in its vicinity. *Phil. Mag.* (ser.4) **46**:26–36.

—— 1875 The rocks of the mining districts of Cornwall and their relation to metalliferous deposits. *Q. J. Geol. Soc. London* **31**:319–345

—— 1879 A contribution to the history of mineral veins. *Q. J. Geol. Soc. London* **35**:390–396

PHILLIPS, R. 1823 Analysis of Uranite from Cornwall. *Ann. Phil.* **21** (n.s. **5**):57–61.

PHILLIPS, W. 1811 A Description of the Red Oxyd of Copper, the production of Cornwall, and of the varieties in the form of its crystal, with observations on the Lodes which principally produced it; and on the Crystallization of the Arseniated Iron. *Trans. Geol. Soc. London* **1**:23–37 + figs. (cuprite, Wheal Gorland, pharmacosiderite)

—— 1814 On the Veins of Cornwall. *Trans. Geol. Soc. London* **2**:110–160

—— 1814 A Description of the Oxyd of Tin, the production of Cornwall; [etc.]: to which is added, a series of its crystalline forms and varieties. *Trans. Geol. Soc. London* **2**:336–376 + Pl.15–26

—— 1816 On the Oxyd of Uranium, the production of Cornwall, together with a description and series of its crystalline forms. *Trans. Geol. Soc. London* **3**:112–120 + Pl.5–7

—— 1817 *An Elementary Introduction to Mineralogy*. London. (2nd ed. 1819; the 3rd ed., 1823, is the one most usually met)

—— 1827 Remarks on the crystalline form of haytorite. *Phil. Mag.* (n.s.) **1**:40–43.

PHIPSON, T. L. 1862 On the argentiferous gossan of Cornwall. *Chem. News* **6**:205–206

PIKE, J. R. 1860 *Britain's metal mines: a complete guide to their laws, usages, localities and statistics*. London.

PIPER, L. P. S. 1974 A short history of Camborne School of Mines. *J. Trevithick Soc.* no.2, pp.9–44

PISANI, F. 1864 Sur quelques nouveaux mineraux de Cornouailles. *C. R. Séances Acad. Sci. Paris* **59**:912–913

POLKINGHORNE, J. P. R. [1951] Bridford Baryte Mine. *TRGSC* **18**:240–254

—— 1957 Tungsten in Cornwall. *TRGSC* **18**:510–525

[POLSUE, J.] 1867–1872 *A Complete Parochial History of the County of Cornwall*; [etc.] 4 vols. Lake, Truro 1867/68/70/72 (usually known as *Lake's Parochial History*, but *see* Boase 1890:1)

POLWHELE, Revd R. 1793–1806 *The History of Devonshire*. Exeter & London 1797 (vol.1); 1793 (vol.2); 1806 (vol.3). (an ambitious plan for vol.1 – see Contents, pp.v-xii – was never completed; vol.1, ch.4, sect.V, p.57–72 concerns 'native fossils', i.e. minerals). (Facsimile reprint, with J. Cary's map (1811), introduction by A. L. Rowse, and ms. index by J. Davidson, by Kohler & Coombes, Dorking 1977).

—— 1803–1808 *The History of Cornwall*: [etc.] 7 vols., complex publication details. Facsimile reprint in 3 bound volumes, with J. Cary's map and introduction by A. L. Rowse, by Kohler & Coombes, Dorking 1978: vols.1 (1803) & 2 (2nd ed., 1816); vols.3 (1803), Supplement (1804, by 'The Historian of Manchester' [J. Whitaker]), & 4 (1806); vols.5 (1806), 6 (1808), & 7 (1806). (Almost nothing on specific minerals, but sections on mines in vols. 1, 3, & 4)

POOL, P. A. S. 1966 William Borlase, the Scholar and the Man. *J. R. Inst. Cornwall* (n.s.) **5**:120–172.

—— 1974 *The History of the Town and Borough of Penzance*. Penzance.

—— 1986 *William Borlase*. Royal Institution of Cornwall, Truro.

POPE, A. *see* Mack, 1969; Searle, 1745

PORTER, R. [S.] 1977 *The Making of Geology / Earth Sciences in Britain 1660–1815*. Cambridge.

—— 1979 John Woodward: 'A Droll Sort of Philosopher'. *Geol. Mag.* **116**:335–343

—— *see* Jordanova 1979

POTTS, R. A. J. 1963 Early Cornish Printers, 1740–1850. *J. R. Inst. Cornwall* (n.s.) **4**(3):264–325.

POWER, G. M. 1968 Chemical variation in tourmalines from South-west England. *Mineral. Mag.* **36**:1078–1089

PRIOR, G. T. 1890 On zinc sulphide replacing stibnite and orpiment; analysis of stephanite and polybasite. *Mineral. Mag.* **9**:9–15

PROVIS, J. 1875 On the lead ores of Cornwall. *Rep. Miners' Assn Cornwall & Devon* (1874), pp.70–77.

PRYCE, W. 1778 *Mineralogia Cornubiensis*; a Treatise on Minerals, Mines, and Mining: [etc.]. London. (facsimile reprint, Barton, Truro 1972)

PUNNETT, H. M. 1859 On some peculiar deposits of Tin in St. Aubyn and Grylls Mine. *TRGSC* **7**:379–380

RAISTRICK, A. (ed.) 1967 *The Hatchett Diary* / A tour through the counties of England and Scotland in 1796 visiting their mines and manufactories. Barton, Truro. (Charles Hatchett accompanied Maton, *q.v.*, on his 1794 tour)

RANKIN, A. H. & ALDERTON, D. H. M. 1985 Chemistry and evolution of hydrothermal fluids associated with the granites of south-west England. In: Halls (org. chairman) 1985:287–300

RASHLEIGH, P. 1797, 1802 *Specimens of British Minerals selected from The Cabinet of Philip Rashleigh* [etc.] with General Descriptions of Each Article. 2 parts. London 1797, 1802. (original coloured plates and ms captions for part 1 in the Mineral Library, British Museum (Natural History)).

—— ms. catalogue (264pp.) of R. collection; also later (1814) cat., by A. Aikin; at R. Instn. Cornwall, County Museum, Truro. (authorised xerographic copies in Mineral Library, BM(NH)).

—— large collection of letters to R. in the Rashleigh papers at the County Record Office, Truro (extracts prepd. by H. L. Douch, 1980); some letters from R., to J. Hawkins, in the Mineral Library, BM(NH)

—— for biography, *see* Russell 1952

RASPE, R. E. 1785 Sur l'Analyse chymique de quelques minéraux remarquables. Lettre de M. Raspe adressée à M. le Conseiller de Collèges Pallas & lue à l'Académie le 23 May. 1785. *Nova Acta Acad. Sci. Imp. Petropolitanae* 1788, Tom.III (for 1785), pp.63–67 (written from 'Entral bey Mamborn [*recte* Camborn] in Cornwallis den 5 Martz 1785'; last sentence, p.67, refers to qual. comp. of stannite, no loc.) (loc., *see* Klaproth 1787).

—— for biography, *see* Carswell 1950

RAYMENT, B. D., DAVIS, G. R. & WILLSON, J. D. 1971 Controls to mineralization at Wheal Jane, Cornwall. *Trans. Inst. Min. Metall.* (Sect.B) **80**:B224–B237

REDDING, C. 1842 *An Illustrated Itinerary of the County of Cornwall*. London.

RICHARDSON, J. B. 1974 *Metal Mining*. Lane, London. (non-ferrous in Britain)

RICKWOOD, P. C. 1981 The largest crystals. *Amer. Mineral.* **66**:885–907 (p.889: BM 42222 Herodsfoot bournonite)

RIDING, R. 1974 Model of the Hercynian Foldbelt. *Earth & Planetary Sci. Letters* **24**:125–135.

RISDON, T. 1811 *The Chorographical Description or Survey of the County of Devon*. London 1811 (Risdon *d.* 1640; 1st ed. pubd. 1714; the present enlarged ed. reprinted 1970)

ROBERTSON, T. & DINES, H. G. 1929 The South Terras radium deposit, Cornwall. *Mining Mag.* **12**:147–153

ROBSON, J. 1945–1948 Geology of Carn Brea. *TRGSC* **17**:208–221; The geology of the St. Ives district, **17**:272–283; The geology of the Land's End peninsula, **17**:427–454

—— [1949] Cornish Mineral Index. *TRGSC* **17**:455–475 (for 'amendments', *see* **18**:406–410)

—— 1953–54. The Cornish greenstones. *TRGSC* **18**:475–492

—— & NANCE, R. M. [1955] Geological Terms used in S.W. England. *TRGSC* **19**:33–41

ROEBUCK, W. R. 1876 *Observations on the iron mines on the Great Perran Lode, with special reference to the spathic iron ores of Cornwall*. (Privately printed and sold by W.P. Collins, Gt. Portland Street, London). 23pp. with a plate of a longitudinal section showing the workings and provings of the iron mines in the Great Perran Lode in Cornwall.

ROGERS, C. 1968 *A Collectors' Guide to Minerals, Rocks, and Gemstones in Cornwall and Devon*. Truro. (ppb.).

ROWE, J. 1953 *Cornwall in the Age of the Industrial Revolution*. Liverpool Univ. Press.

—— 1974 *The Hard-Rock Men* / Cornish Immigrants and the North American Mining Frontier. Liverpool Univ. Press.

RULE, J. 1850 Notice of the discovery of several Rounded Stones in the Lode at South Wheal Frances Mine. *TRGSC* **7**:161–163

RUNDELL, W. W. 1865 Notice of certain peculiar circumstances in Gwinear Consols and Wheal Seton Mines. *TRGSC* **7**:37–39

RUSSELL, A. [E. I. M.] 1910 On the occurrence of the rare mineral Carminite in Cornwall. *Mineral. Mag.* **15**:285–287

—— 1910 Notes on the occurrence of zeolites in Cornwall and Devon. *Mineral. Mag.* **15**:377–384

—— 1911 On the occurrence of Phenacite in Cornwall. *Mineral. Mag.* **16**:55–62

—— 1912 Prehnite from the Lizard district, Cornwall. *Mineral. Mag.* **16**:217–218

—— 1913 An account of the minerals found in the Virtuous Lady mine, near Tavistock. *Mineral. Mag.* **17**:1–14

—— 1913 Notes on the occurrence of Bertrandite at some new localities in Cornwall. *Mineral. Mag.* **17**:15–21

—— 1920 On the occurrence of phenacite and scheelite at Wheal Cock, St. Just, Cornwall. *Mineral. Mag.* **19**:19–22

—— 1920 On the occurrence of cotunnite, anglesite, leadhillite and galena on fused lead from the wreck of the fireship 'Firebrand' in Falmouth Harbour, Cornwall. *Mineral. Mag.* **19**:64–68

—— 1924 Topaz from Cornwall, with an account of its localities. *Mineral. Mag.* **20**:221–236

—— 1924 A notice of the occurrence of native arsenic in Cornwall; [etc.] *Mineral. Mag.* **20**:299

—— 1926 Report on Southern Excursion [of Min. Soc. Jubilee Celebration; many notes on Cornwall and Devon localities visited]. *Mineral. Mag.* **21**:124–128

—— 1927 On laurionite and associated minerals from Cornwall. With a note on paralaurionite by A. Hutchinson. *Mineral. Mag.* **21**:221–228

—— 1927 On the occurrence of the rare mineral nadorite in Cornwall, [etc.]. *Mineral. Mag.* **21**:272–274

—— 1929 On the occurrence of native gold at Hope's Nose, Torquay, Devonshire. *Mineral. Mag.* **22**:159–162

—— 1938 Note on the occurrence [of russellite] and the accompanying minerals. *see* Hey & Bannister 1938

—— 1944 Notes on some minerals either new or rare to Britain. *Mineral. Mag.* **27**:1–10 (gold, realgar, arsenolite, dundasite)

—— 1946 On rhodonite and tephroite from Treburland manganese mine, Altarnun, Cornwall; and on rhodonite from other localities in Cornwall and Devonshire. *Mineral. Mag.* **27**:221–235

—— 1948 *Guide to excursion A4 Devon and Cornwall (mineralogical)*. International Geol. Congress XVIII session, Great Britain.

—— 1948 On rashleighite, a new mineral from Cornwall, intermediate between turquoise and chalcosiderite. *Mineral. Mag.* **28**:353–358

—— 1949 The Wherry mine, Penzance, its history and its mineral productions. *Mineral. Mag.* **28**:517–533

—— 1949 An account of the antimony mines of Great Britain and Ireland, with lists of the minerals found therein. *Mineral. Mag.* **29**:lix (unpublished ms. in Min. Dept. Library, BM(NH))

—— 1949 John Henry Heuland. *Mineral. Mag.* **29**:395–405

—— & VINCENT, E. A. 1952 On the occurrence of varlamoffite (partially hydrated stannic oxide) in Cornwall. *Mineral. Mag.* **29**:817–826

—— 1952 On the occurrence of turquoise in Cornwall. With a chemical analysis by E. A. Vincent. *Mineral. Mag.* **29**:909–912

—— 1952 Philip Rashleigh of Menabilly, Cornwall, and his mineral collection. *J. R. Inst. Cornwall* (n.s.) **1**:96–118.

—— 1954 John Hawkins, FGS, FRHS, FRS, a distinguished Cornishman and early mining geologist. *J. R. Inst. Cornwall* (n.s.) **2**:98–106

—— 1955 The Rev. William Gregor (1761–1817), discoverer of titanium. *Mineral. Mag.* **30**:617–624

—— & CLARINGBULL, G. F. 1955 Ceruleite from Wheal Gorland, Gwennap, Cornwall. *Mineral. Mag.* **31**:xlv (notice only)

—— various mss., mainly on individual minerals, and notes on some collectors and dealers, in Mineral Library, BM(NH)

—— for memorials, *see* Kingsbury 1966

RUST, S. A. 1981 Wheal Gorland. Part 1 – The Mine. *Mineral Kingdom* **1**(2):10–18; Part 2 – The Minerals. *Mineral Realm* **1**(3):6–19, also **1**(6):25. (useful account & sketch plans/ maps, refs limited; note change of journal's title).

—— 1982 Silver Minerals in Cornwall & Devon. *Mineral Realm* **1**(4):20–28

—— 1982 Connellite. *Mineral Realm* **1**(5):31–34.

—— 1982 Wheal Jane. *Mineral Realm* **2**(1):9–20; Mount Wellington mine, **2**(2):9–10.

—— 1983 Wheal Wrey and Ludcott United. *Mineral Realm* **2**(6):10–24

SAINT-FOND, B. F. [de] 1799 *Travels in England, Scotland, and the Hebrides*; [etc.] London. (translation [anon] of the 1797 French edition)

SALMON, H. C. 1862–1864 Illustrated Notes on Prominent Mines. *Mining & Smelting Mag.* **1**:42–49, 314–319, 384–388; **2**:14–18, 74–87, 140–148, 211–223 (incl. Herodsfoot mine and plan). The Seaton Mining District, **2**:277–284, 332–339. East Caradon and 'Sensation' Mines, **2**:339–344. The Condurrow District, **3**:82–89. The St. Ives and Lelant Tin-Mining District, Cornwall, **3**:138–148. The Chiverton Mines, **5**:269–276

—— 1864 The Mines and Mining Operations of Cornwall. *Mining & Smelting Mag.* **5**:257–261, 328–332 (not continued in vol.6, and publication appears to have ceased thereafter).

—— (ed.) 1864 Abstract of article 'The Tin Mines of Cornwall' (Anon., in the *West Briton*). *Mining & Smelting Mag.* **6**:291–293

SANDELL, E. B., HEY M. H. & McCONNELL, D. 1939 The composition of francolite. *Mineral. Mag.* **25**:395–401

SAWKINS, F. J. 1966 Preliminary fluid inclusion studies of the mineralization associated with the Hercynian granites of southwest England. *Trans. Inst. Min. Metall.* (Sect.B) **75**:B109–B112

SCHMITZ, C. J. 1980 *The Teign Valley silver-lead mines 1806–1880*. Northern Mine Research Soc., British Mining No.15 (monograph, 2nd impression 1980; 1st impression, 1973).

—— 1983 Cornish mine labour and the Royal Commission of 1864. *J. Trevithick Soc.* no. 10, pp.30–45

SCRIVENER, R. C., COOPER, B. V., & BAKER, O. A. 1977 Some notes on the mineralization of the Dartmoor granite. *Proc. Ussher Soc.* **4**:7–10

—— & BENNETT, M. J. 1980 Ore genesis and controls of mineralisation in the Upper Palaeozoic rocks of North Devon. *Proc. Ussher Soc.* **5**:54–58

——, SHEPHERD, T. J., & GARRIOCH, N. 1986 Ore genesis at Wheal Pendarves and South Crofty mine, Cornwall – a preliminary fluid inclusion study. *Proc. Ussher Soc.* **6**:412–416.

SCRIVENOR, J. B. 1903 The granite and greisen of Cligga Head (Western Cornwall). *Q. J. Geol. Soc. London* **9**:142–159

SEAGER, A. F. [1969] Mineralisation and Paragenesis at Dean Quarry, The Lizard, Cornwall. *TRGSC* **20**:97–113

—— [1969] Zeolites and other minerals from Dean quarry, the Lizard, Cornwall. *Mineral. Mag.* **37**:147–148

SEARLE, J. 1745 *A Plan of Mr. Pope's Garden, as it was left at his death: with a plan . . . of the Grotto . . . with an account of all the gems, minerals, spars, and ores* [etc.]. London.

SELLECK, A. D. 1978 *Cookworthy 1705–80 and his circle*. Plymouth 1978 (cover title: *Cookworthy / 'a man of no common clay'*).

SELWOOD, E. B. & THOMAS, J. M. 1986 Upper Palaeozoic successions and nappe structures in north Cornwall. *J. Geol. Soc. London* **143**:75–82

SEMMONS, W. 1878 Notes on some silicates of copper, with remarks on the chrysocolla group. *Mineral. Mag.* **2**:197–205

—— 1885 Notes on a recent discovery of 'connellite'. *Mineral. Mag.* **6**:160–163

SEYMOUR, G. (Jun.) 1878 On the occurrence of tin in an elvan course at Wheal Jennings. *TRGSC* **9**:185–195

—— 1977 *The man-machine and other sketches*. Scenes from Cornish Tin Mining in the 1870s. Mining Journal, London. (originals drawn in 1877).

SHACKLETON, R. M., RIES, A. C. & COWARD, M. P. 1982 An interpretation of the Variscan structures in SW England. *J. Geol. Soc. London* **139**:533–541

SHAMBROOK, H. R. [1982] *The Caradon and Phoenix mining area*. Northern Mine Research Soc., Sheffield. British Mining No.20. (monograph; 2nd ed. 1986, publ. by the author).

—— 1982 The Devon Great Consolidated Copper Mining Company. *J. Trevithick Soc.* no.9, pp.62–68

SHEPPARD, S. M. F. 1977 The Cornubian Batholith, south west England: D/H and $^{18}O/^{16}O$ studies of kaolinite and other alteration minerals. *J. Geol. Soc. London* **133**:573–591

[SIMMONDS, A.] 1942 The Founders. The Rt. Hon. Charles Greville, F.R.S., F.L.S. (1749–1809). *J. R. Horticult. Soc.* **67**:219–232

SIMPKINS, D. M. 1974 Biographical sketch of James Sowerby written by his son, James de Carle Sowerby. *J. Soc. Biblphy Nat. Hist.* **6**(6):402–415

SIMPSON, P. R., BROWN, G. C., PLANT, J. & OSTLE, D. 1979 Uranium mineralization and granite magmatism in the British Isles. *Phil. Trans. R. Soc. London* **A291**:385–412

SMALE, C. V. 1977 The Cornish china stone industry. *Trans. Cornish Inst. Eng.* **31**:1–13

SMITH, E. 1817 On the Stream Works of Pentowan. *Trans. Geol. Soc. London* **4**(2):404–409

SMYTH, W. W. 1859. On the iron-ores of Exmoor. *Q. J. Geol. Soc. London* **15**:105–109

—— 1865 On the iron mines of Perran. *TRGSC* **7**:332–335

—— 1878 On the occurrence of metallic ores with garnet rock. Note illustrating a series of copper ores from Belstone Consols. *TRGSC* **9**:38–45

—— 1887 The Duchy Peru Lode, Perranzabuloe. *TRGSC* **10**:120–131

SOLLY, R. H. 1885 Notes on minerals from Cornwall and Devon. *Mineral. Mag.* **6**:202–211. (Holmbush fluorite; Belstone minerals; long note on axinite locs.).

—— 1886 Francolite, a variety of apatite from Levant Mine, St. Just, Cornwall. *Mineral. Mag.* **7**:57–58.

—— 1887 Apatite from the Levant Mine, Cornwall. *TRGSC* **10**:240–244

—— 1891 Cassiterite, 'Sparable Tin', from Cornwall. *Mineral. Mag.* **9**:199–208 (cryst., many locs)

SOWERBY, J. 1804–1817 *British Mineralogy: or Coloured Figures intended to elucidate The Mineralogy of Great Britain*. 5 vols. London 1804, 1806, 1809, 1811, & 1817.

—— 1811 *A short Catalogue of British Minerals, according to a new arrangement*. London. (footnote p.23: claims orig. discovery of 'hydrargillite' [= wavellite] by I. Hill, before 1785).

—— for biogr., etc., *see* Simpkins 1974; Cleevely 1974, 1976

SPARGO, T. 1865 *The Mines of Cornwall and Devon*: statistics and observations, illustrated by maps etc . . . London. (Cornwall section only, repr. in 6 ppb. parts, Barton, Truro 1959–1961).

—— 1868 *The Mines of Cornwall and Devon*: etc. Head, London. (an enlarged ed., 200pp.) (other eds. in 1859, 1860, 1862, 1864, 1867, 1870)

—— 1872 *The Tin Mines of Cornwall and Devon, their present position and prospects*. London.

SPENCER, L. J. 1921–1947 Biographical notices of mineralogists recently deceased. *Mineral. Mag.* **19**:240–262; **20**:252–275; **21**:229–257; **22**:387–412; **23**:337–366; **24**:277–306; **25**:283–304; **28**:175–229

—— 1958 Third supplementary list of British minerals. *Mineral. Mag.* **31**:787–806 (1st & 2nd lists, reprinted, **31**:807–810) (see Greg & Lettsom, 1858/1977)

STANIER, P. 1985 Granite-working in the Cheesewring district of Bodmin Moor, Cornwall. *J. Trevithick Soc.* no.12, pp.36–51

—— 1986 John Freeman and the Cornish granite industry. *J. Trevithick Soc.* no.13, pp.7–35

STEER, F. W. 1966 *The Letters of John Hawkins and Samuel and Daniel Lysons 1812–1830*. Chichester.

STEPHENS, F. J. 1893 Notes on the Marazion and Perranuthnoe mining districts. *Rep. R. Cornwall Poly. Soc.* (1893), pp.113–120

—— 1895 On a Supposed Resemblance between the Occurrence of Native Copper in the Lake Superior and Lizard Areas. *TRGSC* **11**:680–683

—— 1896 Some mining and geological notes on the Crowan and Gwinear district. *Rep. R. Cornwall Poly. Soc.* **64**:44–55

—— 1899 Recent Discoveries of Gold in West Cornwall. *TRGSC* **12**:241–257

—— 1900 Alluvial deposits in the lower portions of the Red River Valley, near Camborne. *TRGSC* **12**:324–335

—— 1918 Notes on the Mining District between Camborne and Redruth North of the Main Road. *TRGSC* **15**:286–307 (much anecdotal material)

—— 1928 Notes on ancient mining in Cornwall. *Rep. R. Cornwall Poly. Soc.* (1928), pp.162–171

—— 1932 The Ancient Mining Districts of Cornwall. Notes on the Geology, Minerals, and Mines of the Liskeard District within an area of five to six miles. *Rep. R. Cornwall Poly. Soc.* (n.s.) **7**:159–178.

—— 1937 The ancient mining districts of Cornwall. North of Truro and parts of Newlyn East, Cubert, Perranzabuloe and St. Allen parishes. *Rep. R. Cornwall Poly. Soc.* (n.s.) **9**(1):70–81

—— 1938 The ancient mining districts of Cornwall. Illogan, St. Agnes and Western Perranzabuloe parishes. *Rep. R. Cornwall Poly. Soc.* (n.s.) **9**(2):33–56.

—— 1939 North Gwennap and Illogan. *Mining Mag.* **61**:83–90

—— 1940 The south Gwennap mining district, with a portion of Baldhu and Kea. *Mining Mag.* **62**:9–19

—— 1940 The Wendron mining district with notes on the flat lode and Old Wheal Vor. *Mining Mag.* **63**:233–240

STEPHENS, H. 1871 Mineral phenomena of Huel Rose. *Rep. R. Cornwall Poly. Soc.* **39**:77–80 (also *Rep. Miners' Assn Cornwall & Devon* (1871), pp.47–49)

STOCKDALE, F. W. L. 1824 *Excursions in the County of Cornwall* [etc.] London. (reprinted Barton, Truro 1972, with cover title *Excursions Through Cornwall*).

STONE, M. 1961 Genesis of the granites of SW England. *Proc. 4th*

Conf. Geol. Geomorph. SW England. R. Geol. Soc. Cornwall, Penzance, pp.5–7.
—— 1968 A study of the Praa Sands elvan and its bearing on the origin of elvans. *Proc. Ussher Soc.* **2**:37–42
—— 1969 Nature and origin of banding in the granitic sheets, Tremearne, Porthleven, Cornwall. *Geol. Mag.* **106**:142–158
—— 1975 Structure and petrology of the Tregonning-Godolphin Granite, Cornwall. *Proc. Geol. Assn* **86**:155–170
—— 1984 Textural evolution of lithium mica granites in the Cornubian batholith. *Proc. Geol. Assn* **95**:28–41
—— & AUSTIN, W. G. C. 1961 The metasomatic origin of the potash feldspar megacrysts in the granites of southwest England. *J. Geol.* **69**:464–472
—— & GEORGE, M. C. 1978 Amblygonite in leucogranites of the Tregonning-Godolphin Granite, Cornwall. *Mineral. Mag.* **42**:151–152
—— & —— 1983 Some phosphate minerals at the Megiliggar Rocks, Cornwall. *Proc. Ussher Soc.* **5**:428–431
—— & EXLEY, C. S. 1985 High heat production granites of southwest England and their associated mineralization: a review. In: Halls (org. chairman) 1985:571–593
STRONG, H. W. 1890 A contribution to the Commercial History of Devonshire. *Trans. Devon. Assn* **22**:129–137
STYLES, M. T. & KIRBY, G. A. 1980 New investigation of the Lizard complex, Cornwall, England and a discussion of an ophiolite model. *Proc. Internat. Ophiolite Symposium*. Cyprus Geol. Surv. Dept., 1980:517–526
SVEDENSTIERNA, E. T. 1804 *see* Dellow 1973
SYMONDS, W. S. 1872 *Records of the Rocks*; or, notes on the geology, natural history, and antiquities of north and south Wales, Devon, & Cornwall. Murray, London.
SYMONS, B. 1864 Geological map of the Crowan and Wheal Abraham Mining Districts
—— 1884 *A Sketch of the Geology of Cornwall*, including a Brief Description of the Mining Districts and the ores produced in them. London. (reprint of pp.187–235 in Symons, R., 1884).
SYMONS, R. 1845 Geological and mining map of the Parish of Gwennap, scale 6":1 mile.
—— 1870 Map of the Camborne, Illogan, Redruth and Gwennap Mines, scale 5":1 mile. Robert Symons & Son, Mineral Surveyors, Truro.
—— 1878 On Carclaze tin and china clay pit. *J. R. Inst. Cornwall* **6**:140–143
—— 1884 *A Geographical Dictionary, or Gazetteer of the County of Cornwall*; with a treatise on the Geology of Cornwall, (and map), by Brenton Symons. Penzance.
TANGYE, M. 1981 *Carn Brea* / (Rocky Hill). Dyllansow Truran, Redruth. (ppb.; brief history and guide, little about mines).
—— 1984 *Tehidy and the Bassets*. Dyllansow Truran, Redruth. (ppb.; almost nothing about the family mining interests).
TAYLOR, C. D. 1873 Description of the tin stream works in Restronguet Creek, near Truro. *Proc. Inst. Mech. Eng.* 1873, pp.155–166 + Pl. 58, 59.
TAYLOR, J. 1799 Sketch of the History of Mining in Devon and Cornwall. *Phil. Mag.* **5**:357–365.
—— 1814 On the Economy of the Mines of Cornwall and Devon. *Trans. Geol. Soc. London* **2**:309–327
—— 1817 Description of the Tunnel of the Tavistock Canal, through Morwel Down, in the County of Devon. *Trans. Geol. Soc. London* **4**(2):146–155
—— (ed.) 1829 *Records of mining*. Part 1. John Murray, London. (174pp., 17 pl.; also, tables exhibiting the quantities of copper, tin &c. produced in Great Britain, 1799–1828. pp.167–174; no other parts publ.). (repr. Mining Facsimiles, Sheffield 1986).
—— 1834 Account of the depths of mines. *Rep. Brit. Assn* (1833), pp.427–430
—— *see also* Burt 1969; biography, Burt 1977
TAYLOR, K. 1958 Coffinite in Cornwall. *Nature* **181**:363
TAYLOR, R. 1846 On the relative position of the yellow and vitreous sulphurets of copper in the lode of Pembroke Mine. *TRGSC* **6**:99–100
TAYLOR, R. G. 1965 The throw of the Great Cross Course in the Camborne-Redruth mining district, Cornwall. *Trans. Inst. Min. Metall.* **74**:529–543
—— 1966 Distribution and deposition of cassiterite at South Crofty Mine, Cornwall. *Trans. Inst. Min. Metall.* (Sect.B) **75**:B35–B49
—— 1969 Influence of early quartz-felspar veins on cassiterite distribution at South Crofty Mine, Cornwall. *Trans. Inst. Min. Metall.* (Sect.B) **78**:B72–B85
TEALL, J. J. H. 1887 On granite containing andalusite from the Cheesewring, Cornwall. *Mineral. Mag.* **7**:161–163
—— 1890 Metamorphism in the Hartz and West of England. *TRGSC* **11**:221–238
TERRELL, E. 1920 The Hemerdon wolfram-tin mine. *Mining Mag.* **22**:75–87
TEW, D. H. 1981 Man Engines in Cornwall. *J. Trevithick Soc.* no.8, pp.47–53
THOMAS, C. 1855 Remarks on mining in Cornwall and Devon. *Rep. R. Cornwall Poly. Soc.* **22**:28–35
—— 1859 *Remarks on the geology of Cornwall and Devon* in connexion with the deposits of metallic ores and on the bearings of the productive loads. Given in two lectures, with lithographic illustrations, together with additional remarks on the same subject. Jas. Tregaskis, Ticketing paper office, Redruth.
—— 1871 *Mining Fields of the West*: being a Practical Exposition of the Principal Mines and Mining Districts in Cornwall and Devon. (2nd ed. 1871; ppb. reprint Barton, Truro 1967).
THOMAS, H. 1896 *Cornish Mining Interviews* / copiously illustrated, with portraits of celebrities and photographs with mining scenes at home and abroad, including: Underground Scenes by J. C. Burrow. Camborne Printing & Stationery Co. Ltd.
—— (ed.) [1922] *One and All to Save Cornish Tin Mines* / The Tragedy of an Ancient Staple Industry. (publ. by ??) (reprints of newspaper articles, etc.).
THOMAS, J. 1870 Description of the operations at Dolcoath Mine. *J. R. Inst. Cornwall* **3**:191–197
THOMAS, R. 1819 *Report on a survey of the mining district of Cornwall from Chasewater to Camborne*. John Cary, London. (77pp., with: A geological map of the mining district of Cornwall between Camborne and Chasewater showing the lodes, cross-courses and adits etc., Geological sections in illustration of the map of the mining district of Cornwall, and, Geological view of the mining district of Cornwall corresponding with the map from Chasewater to Camborne).
—— 1836 Geological survey of the Carn Menelis District, Cornwall. *Mining Rev.* **3**:263–280
—— 1837 Geological particulars of parts of the cliffs in the Land's End district, Cornwall. *Mining Rev.* (n.s.) **1**:234–242.
THOMAS, R. H. 1886 Some observations on the 'Great Flat Lode' in Wheal Uny Mine, near Redruth. *Rep. R. Cornwall Poly. Soc.* **54**:184–188
THOMAS, W. 1925 Some Cornish Mines now under Water. *Mining Mag.* **33**:16–20
THOMAS, W. R. 1907 Electrically driven centrifugal pumping plant at Tywarnhayle Mine. *Trans. Inst. Min. Metall.* **16**:206–230
THORNE, M. [G.] 1981 Trends in the hard rock mining industry of

South West England. *British Geologist 1981* 7(4):92–98. (repr. in *J. Camborne Sch. Mines* **81**:45–52)

THORNE, M. G., & EDWARDS, R. P. 1985 Recent Advances in Concepts of Ore Genesis in South West England. *TRGSC* **21**(3):113–152

TILLEY, C. E. 1923 The petrology of the metamorphosed rocks of the Start area (South Devon). *Q. J. Geol. Soc. London* **79**:172–204

—— 1925 Petrographical notes on some chloritoid rocks. II. Chloritoid phyllites of the Tintagel area, North Cornwall. *Geol. Mag.* **62**:314–318

—— 1935 Metasomatism associated with the greenstone-hornfelses of Kenidjack and Botallack, Cornwall. *Mineral. Mag.* **24**:181–202

—— 1946 Bustamite from Treburland manganese mine, Cornwall, and its paragenesis. *Mineral. Mag.* **27**:236–241

—— 1873 Particulars of a thermal spring at Wheal Seton in the Parish of Camborne, with a comparative table of analyses of similar springs in the United, Balleswidden, Botallack and Crown Mines. *Rep. Miners' Assn Cornwall & Devon* (1872–73), pp.53–56

TILLOCH, A. (ed.) 1801 *Phil. Mag.* **11**:381–382 (Mining in Devon and Cornwall – new discoveries; list of minerals found 'of late').

TODD, A. C. [1960] Origins of the Royal Geological Society of Cornwall. *TRGSC* **19**:179–184.

—— 1964 The Royal Geological Society of Cornwall. Its Origins and History, based upon its official Minutes, Reports and Transactions. In: Hosking & Shrimpton, 1964:1–23.

—— 1967 *The Cornish Miner in America* / the contributions to the mining history of the United States by emigrant Cornish miners – the men called Cousin Jacks. Barton, Truro / Clark, Glendale.

—— 1967 *Beyond the Blaze / A biography of Davies Gilbert.* Truro.

—— 1977 *The search for Silver: Cornish Miners in Mexico, 1824–1947.* Lodenek Press, Padstow.

—— & LAWS, P. 1972 *The Industrial Archaeology of Cornwall.* David & Charles, Newton Abbot.

TOLL, R. W. 1938 The arsenic industry in the Tavistock district of Devon. *Sands, Clays & Minerals* **3**:224–227.

TOMBS, J. M. C. 1977 Concealed granite in Cornwall : its shape from its gravitational effect. *Trans. Inst. Min. Metall.* (Sect.B) **86**:B93–B95

TOOKE, A. W. 1836 The mineral topography of Great Britain (Cornwall). *Mining Rev.* **3**:253–263

TORRENS, H. 1977 (note on the St Aubyn coll.). *Geol. Curators' Grp Newsletter* **1**(9):455–456.

—— 1985 Early Collecting in the Field of Geology. Ch.24 In: Impey & Macgregor (eds.) 1985:204–213.

[TRADESCANT, J.] 1656 *Musaeum Tradescantianum*: or, a Collection of Rarities. Preserved at South-Lambeth neer London By John Tradescant. London. (facsim. repr. by Hewett, London, 1980)

TRECHMANN, C. O. 1885 Connellite from Cornwall. *Mineral. Mag.* **6**:171

TREDINNICK, R. 1857 *A review of Cornish and Devon mining enterprise, 1850–1856 inclusive.* Thomson & Vincent, London.

—— 1858 *A review of Cornish copper mining enterprise, ... and a detailed account of the Buller and Basset District.* Thomson & Vincent, London 1858. 134pp. + maps.

TREVITHICK, F. 1872 *Life of Richard Trevithick*, with an account of his inventions. 2 vols. London / New York.

TRINICK, G. M. A. 1981 The Tregurtha Downs mines, Marazion, 1700–1965. *J. Trevithick Soc.* no.8, pp.7–25

TRIPE, C. 1827 Observations on a mineral from near Hay Tor. *Phil. Mag.* (n.s.) 1:38–40. (haytorite)

TROUNSON, J. H. 1940 A Cornish wolfram producer. *Mining Mag.* **63**:18–28

—— 1942 The Cornish mineral industry, Part iii, St. Just and St. Ives mining districts. *Mining Mag.* **67**:119–130

—— 1959 Practical Considerations in Developing Old Cornish Mines. In: *The future of non-ferrous mining in Great Britain and Ireland.* Proceedings of a symposium, London 1959, pp.371–382.

—— 1960 *Mineral Areas in Cornwall Worthy of Investigation.* Cornish Mining Development Association (5th ed., 1960)

—— 1968 *Historic Cornish Mining Scenes at Surface.* Barton, Truro.

—— [1980] *Mining in Cornwall 1850–1960.* 2 vols. (On behalf of the Trevithick Society) Moorland Publ. Co. Ltd., Ashbourne, (n.d.) [1980; 1981]

—— 1982 Cornish stacks and engine houses. *J. Trevithick Soc.* no.9, pp.73–84

—— 1984 The Cornish mining industry in the 19th and 20th centuries. *J. Trevithick Soc.* no.11, pp.7–17

—— *see also* under Hosking

TUCKER, D. G. & M. The Story of Wheal Guskus in the Parish of Saint Hilary. *J. Trevithick Soc.* no.1, pp.49–62

TURK, F. A. 1959 Natural History Studies in Cornwall (1700–1900) *J. R. Inst. Cornwall* (n.s.) **3**:229–279.

—— *see also* under Borlase

TURNER, J. 1975 *The Stone Peninsula* / Scenes from the Cornish Landscape. Kimber, London. (general but interesting).

TWEEDY, W. M. 1846 A Description of the Lode at Wheal Coates Mine, in which the Pseudomorphous crystals of Tin occurred. *Rep. R. Inst. Cornwall* 1846, App.II, pp.20–24.

VALE, E. 1966 *The Harveys of Hayle / Engine builders, shipwrights, and merchants of Cornwall.* Barton, Truro. (for catalogue reprint, *see* Harvey & Co.).

VALLANCE, T. G. 1981 The start of government science in Australia: A. W. H. Humphrey, His Majesty's Mineralogist in New South Wales, 1803–12. *Proc. Linn. Soc. N.S.W.* **105**:107–146.

VANCOUVER, C. 1808 *General view of the agriculture of the county of Devon.* Richard Phillips, London, for the Board of Agriculture.

VAN MARCKE DE LUMMEN, G. & VERKAEREN, J. 1985 Mineralogical observations and genetic considerations relating to skarn formation at Botallack, Cornwall, England. In: Halls (org. chairman) 1985:535–547.

VIVIAN, J. 1972 *Tales of the Cornish miners.* Tor Mark, Truro.

VOKES, F. M. & JEFFERY, W. G. 1955 The geology of New Consols Mine, Cornwall. *Bull. Inst. Min. Metall.* **64**:141–164 + 4 figs.

VON OEYNHAUSEN, C. & VON DECHEN, H. 1829 On the junction of the granite and the killas rocks of Cornwall. *Phil. Mag.* (ser.2) **5**:161–170, 241–247.

WARD, G. R. 1983 Bertrandite from Hingston Down Quarry, Calstock, Cornwall. *Proc. Ussher Soc.* **5**:485

WARNER, R. 1809 *A Tour through Cornwall in the Autumn of 1808.* London.

WATSON, J. V., FOWLER, M. B., PLANT, J. A. & SIMPSON, P. R. 1984 Variscan-Caledonian comparisons: late orogenic granites. *Proc. Ussher Soc.* **6**:2–12.

WATSON, J. Y. 1843 *A compendium of British mining* with statistical notices of the principal mines in Cornwall; [etc.]. Printed for private circulation by Munro & Congreve, London. (ppb. repr. in 1960s, Barton, Truro).

—— 1861 *Cornish notes for 'Out-Adventurers'.* Mining Journal (Publishers), London, 31pp. (repr. ppb. Truro Bookshop, Truro, 1961, with cover title *Cornish Mining Notes 1861*)

WEBB, (W.L.?) & GEACH, E. 1862 *The history and progress of mining in the Caradon and Liskeard district.* Williams & Strachan, London. (2nd ed., Effingham Wilson, London 1863)

WEEKS, M. E. & LEICESTER, H. M. 1968 *Discovery of the Elements*. 7th ed., Easton. (esp. biogr. of Charles Hatchett).

WEINDLING, P. J. 1979 Geological controversy and its historiography: the prehistory of the Geological Society of London. In: Jordanova & Porter (eds.) 1979:248–271.

—— 1983 The British Mineralogical Society, 1799–1806. A case-study of science and social improvement. In: Inkster, I. & Morrell, J. *Metropolis and Province / Science in British culture, 1780–1850*. London. pp.120–150

WEINER, K. L. & HOCHLEITNER, R. 1986 Steckbrief: Lirokonit. *Lapis* **11**(2):9–11.

WEISS, C. S. 1828 Ueber den Haytorit. *Abh. preuss. Akad. Wiss. Berlin* 1828, pp.63–76

WEISS, S. 1986 Geografie und Geologie von Cornwall. *Lapis* **11**(2):13–15; Der Bergbaubezirk Camborne-Redruth. **11**(2):16–38; Die Bergbaugeschichte Cornwalls. **11**(2):39–41; Die Grubenbezirke St. Just und St. Ives in Cornwall. **11**(5):9–32.

[WHITAKER, J.] 1804 'The Historian of Manchester', *see* Polwhele 1803

WHITAKER, W. 1870 List of works on the geology, mineralogy, and palaeontology of Devonshire. *Rep. Trans. Devon. Assn* **4**:330–352 (supplement, 1872, **5**:404)

—— 1874 List of works on the geology, mineralogy and palaeontology of Cornwall. *J. R. Inst. Cornwall* **5**:61–110

WHITEHEAD, P. J. P. 1973 Some further notes on Jacob Forster (1739–1806), mineral collector and dealer. *Mineral. Mag.* **39**:361–363

WHITLEY, N. 1878 The geology of Penzance Bay and its shores. *TRGSC* **9**:109–113 (+ map)

WHITLEY, REV D. G. 1912 The earliest traces in ancient history of the tin trade of Western Europe. *TRGSC* **13**:515–529

—— 1914 The early tin trade of Cornwall according to Strabo's Geography. *TRGSC* **13**:595–610

—— 1915 The Ictis of Diodorus in the light of modern theories. *TRGSC* **15**:55–70

WHITWORTH, J. S. 1914 The lodes of the St. Agnes district. *Trans. Cornish Inst. Min. Mech. Metall. Eng.* **2**:214–232 + maps.

WILKINSON, W. F. 1895 The history of Holmbush, Redmoor and Kelly Bray. *Mining J.* 1895, pp.34, 62, 90.

WILLIAMS' PERRAN FOUNDRY CO.(publ.) [c.1870] *Illustrated Catalogue of Pumping & Winding Engines, and Other Plant used for Mining Purposes*, [etc.] manufactured by Williams' Perran Foundry Co., Perranarworthal, Cornwall, and . . . . London. (Trevithick Society reprint no.1, 1976; with brief history 1791–1879 by T. R. Harris. compare Harvey & Co. catalogue).

WILLIAMS, F. M. 1871 Recent observations on subterranean temperature in the Clifford Amalgamated Mines. *J. R. Inst. Cornwall* 3:283–285

WILLIAMS, G. 1984 A history of Giew mine. *J. Trevithick Soc.* no.11, pp.60–70 (also unnumbered page between p.56 & p.57)

—— 1986 Bal Du mine or West Wheal Reeth. *J. Trevithick Soc.* no.13, pp.80–83

WILLIAMS, H. V. [1969] *Cornwall's Old Mines*. Tor Mark Press, Truro (ppb.; n.d., ? 1969)

WILLIAMS, J. 1817 Account of some remarkable Disturbances in the Veins of the Mine called Wheal Peever, in Cornwall. *Trans. Geol. Soc. London* **4**(2):139–145

WILLIAMS, J. 1861–1870. *The Cornwall and Devon mining directory*. (1st ed. Hayle 1861; 2nd ed. London 1862; 3rd ed. 1870)

WILLIAMS, J. A. [?1970s] *Cornish Tokens*. Barton, Truro. (ppb., n.d.; pictures of mining tokens pp.41–42).

WILLIAMS, R. H. 1858 Note on the occurrence of nickel and cobalt at St. Austell Consols Mines. *Rep. R. Inst. Cornwall* **39**:32–34 (App. vii).

WILLIAMS, W. H. 1863 On mineral deposits. *Rep. R. Inst. Cornwall* **45**:44–47 (Great Crinnis mine).

WILSON, H. E. 1985 *Down to Earth* / One hundred and fifty years of the British Geological Survey. Scottish Academic Press, Edinburgh & London.

WILSON, I. R. 1972 Wallrock alteration at Geevor Mine. *Proc. Ussher Soc.* **2**:425–434.

—— & FLOYD, P. A. 1974 Distribution of uranium in the Land's End granite and aureole, and various greenstones from Cornwall. *Proc. Ussher Soc.* **3**:166–176

WOLLOXALL, T. 1987 Levant Mine. *Rock Bottom* (U.K. Journal of Mines & Minerals) no.2, pp.24–27.

WOLTERS, J. & A. [1986] Dean Quarry. *Rock Bottom* (U.K. Journal of Mines & Minerals) no.1, pp.20–27. (black & white photos).

WOOD, S. V. 1885 Cornish tin mining, Great Wheal Vor district. *Mining J.* **55**:987.

WOODFIN, R. J. 1972 *The Cornwall Railway* / to its centenary in 1959. Barton, Truro. (unrevised reprint of *The Centenary of the Cornish Railway*, 1960). (esp. ch.1, on earlier lines).

WOODWALL, F. D. 1975 *Steam Engines and Water Wheels* / a pictorial study of some early mining machines. Moorland Publishing Co., Buxton.

WOODWARD, H. B. 1907 *The History of the Geological Society of London*. London.

WOODWARD, J. 1696 *Brief Instructions for Making Observations in All Parts of the World, as Also for Collecting, Preserving, and Sending Over Natural Things*. London. (repr., London 1973)

—— 1728–1729 *An Attempt Towards a Natural History of the Fossils of England; in a Catalogue of the English Fossils in the Collection of J. Woodward, M.D.* [etc.]. London. (Tome I, 1729; Tome II, 1728).

—— biography, etc.: Eyles, 1971; Levine, 1977; Porter, 1979; Torrens, 1985

WORTH, R. H. 1902–1903 The petrography of Dartmoor and its borders, Parts 1 & 2. *Trans. Devon Assn* **34**:496–527; **35**:759–767

WORTH, R. N. 1874 The antiquity of mining in the West of England. *Ann. Rep. & Trans. Plymouth Inst.* **6**:120–140

—— 1886 The rocks and minerals of Cornwall and Devon. *Rep. R. Cornwall Poly. Soc.* **59**:70–88

—— 1887 The igneous and altered rocks of south-west Devon. *Trans. Devon. Assn* **19**:467–497

—— 1890 Contact metamorphism in Devonshire. *Trans. Devon. Assn* **22**:169–184

—— 1893 The age and history of the granites of Devon and Cornwall. *TRGSC* **11**:480–486

WYNNE, W. 1755 *see* Edwards, 1981

ZAGHLOUL, Z. M. 1958 Radioactivity and radioactive accessory minerals of the Land's End granite. In: Proc. 2nd Conf. Geol. Geomorph. SW England. R. Geol. Soc. Cornwall, Penzance. pp.8–9.

Geological Survey of Great Britain:
Memoirs, Reports, etc

(now British Geological Survey; former titles include Institute of Geological Sciences)

(Note: where there are different editions, the earlier may contain mining information omitted in the later.)

General:

[Anon] 1896 *A Handbook to the Museum of Practical Geology*. 5th ed. London. (earlier eds. 1857, 1859, 1867, 1877).

BECHE, H. T. de la 1839 *Report on the Geology of Cornwall, Devon, and West Somerset*. London. (Index, by C. Reid, publ. 1903)

BEER, K. E., BENNETT, M. J., BALL T. K., JONES R. C., & TURTON, K. 1981 Metalliferous mineralization near Lutton, Ivybridge, Devon. Mineral Reconn. Prog. Report no. 41.

BENNETT, M. J., BEER, K. E., JONES, R. C., TURTON, K., ROLLIN, K. E., TOMBS, J. M. C., & PATRICK, D. J. 1981 Mineral investigations near Bodmin, Cornwall. part 3 – the Mulberry and Wheal Prosper area. Mineral Reconn. Prog. Report no. 48.

DINES, H. G. 1956 *The Metalliferous Mining Region of South-West England*. 2 vols. (2nd ed. in preparation)

EDMONDS, E. A., McKEOWN, M. C., & WILLIAMS, M. 1975 (4th ed.) *British Regional Geology: South-West England*. (3rd ed. 1969; 1st ed. 1935)

GOODE, A. J. J. 1973 The mode of intrusion of Cornish elvans. Report 73/7

—— & TAYLOR, R. T. 1980 Intrusive and pneumatolytic breccias in south-west England. Report 80/2

HAWKES, J. R., HARDING, R. R., & DARBYSHIRE, D. P. F. 1975 Petrology and Rb:Sr age of the Brannel, South Crofty, and Wherry elvan dykes, Cornwall. *Bull. Geol. Surv. G.B.* **52**:27–42.

HIGHLEY, D. E. 1984 China Clay. Mineral Resources Consultative Committee, Mineral Dossier no. 26

HUNT, R. 1855–1882. Mineral statistics of the United Kingdom of Great Britain and Ireland for the years 1853 through to 1881 (29 volumes). *Mems. Geol. Surv. Great Britain*. Mining Records.

MacALISTER, D. A. 1907 Total quantity of tin, copper and other minerals produced in Cornwall, particularly with regard to the quantities raised from each parish. Sum. Prog. Geol. Surv. for 1906, pp.132–139.

—— 1921. Total quantity of tin, copper and other minerals produced in Devonshire, particularly with regard to the amounts raised from each parish. Sum. Prog. Geol. Surv. for 1920, pp.96–102.

RUDLER, F. W. 1905 *A Handbook to a Collection of the Minerals of the British Islands*, mostly selected from the Ludlam Collection, in the Museum of Practical Geology, Jermyn Street, London, S.W. London. (also, ms. catalogue of the Ludlam coll.).

SLATER, D. 1973 Tungsten. Mineral Resources Consultative Committee, Mineral Dossier no. 5.

—— 1974 Tin. Mineral Resources Consultative Committee, Mineral Dossier no. 9.

SMYTH, W. W. (with T. REEKS, & F. W. RUDLER) 1864 *A Catalogue of the Mineral Collection in the Museum of Practical Geology*. London.

TAYLOR, J., HARRISON, R. K., & TAYLOR, K. 1966 Structure and mineralization at Roskrow United Mine, Ponsanooth, Cornwall. *Bull. Geol. Surv. G.B.* **25**:33–40.

WILSON, A. C. 1975 An occurrence of sapphire in the Land's End granite, Cornwall. *Bull. Geol. Surv. G.B.* no.52, pp.61–63.

District Memoirs / 1″ Sheet explanations:

Bideford and Lundy Island (Sheets 292 n.s. and others), Geology of. Edmonds, E. A., Williams, B. J., & Taylor R. T. 1979.

Bodmin and St Austell (Sheet 347), The Geology of the Country around. Ussher, W. A. E., Barrow, G., & MacAlister, D. A. E. 1909.

Boscastle and Holsworthy (Sheets 322 and 323 n.s.), The Geology of the Country around. McKeown, M. C., Edmonds, E. A., Williams, M., Freshney, E. C., & Masson Smith, D. J. 1973.

Bude and Bradworthy (Sheets 307 and 308), Geology of the Country around. Freshney, E. C., Edmonds, E. A., Taylor, R. T., & Williams, B. J. 1979.

Dartmoor (Sheet 338), The Geology of. Reid, C., Barrow, G., Sherlock, R. L., MacAlister, D. A., Dewey, H., & Bromehead, C. N. 1912.

Exeter (Sheet 325), The Geology of the Country around. Ussher, W. A. E. 1902.

Falmouth and Truro and the Mining District of Camborne and Redruth (Sheet 352), The Geology of. Hill, J. B. & MacAlister, D. A. 1906.

Ilfracombe and Barnstaple (Sheets 277 and 293 n.s.), Geology of the Country around. Edmonds, E. A., Whittaker, A., & Williams, B. J. 1985.

Ivybridge and Modbury (Sheet 349), The Geology of the Country around. Ussher, W. A. E. 1912.

Land's End District (Sheets 351 and 358), The Geology of the. Reid, C. & Flett, J. S. 1907. (with Mining Appendix by D. A. MacAlister).

Lizard and Meneage (Sheet 359), The Geology of the. Flett, J. S. & Hill, J. B. 1912. (2nd ed. 1946; repr. with add. refs, 1973).

Mevagissey (Sheet 353), The Geology of the Country around. Reid, C. 1907.

Newquay (Sheet 346), The Geology of the Country near. Reid, C. & Scrivenor, J. B. 1906. (with contribs. by Flett, Pollard, & MacAlister).

Newton Abbot (Sheet 339), The Geology of the Country around. Ussher, W. A. E. 1913 (with contribs. by Read, Flett, & MacAlister); new edition, by Selwood, E. B. *et al.*, 1984.

Okehampton (Sheet 324 n.s.), Geology of the Country around. Edmonds, E. A., Wright, J. E., Beer, K. E., Hawkes, J. R., Williams, M., Freshney, E. C., & Fenning, P. J. 1968.

Padstow and Camelford (Sheets 335 and 336), The Geology of the Country around. Reid, C., Barrow, G., & Dewey, H. 1910.

Plymouth and Liskeard (Sheet 348), The Geology of the Country around. Ussher, W. A. E. 1907.

Tavistock and Launceston (Sheet 337), The Geology of the Country around. Reid, C., Barrow, G., Sherlock, R. L., MacAlister, D. A., & Dewey, H. 1911.

Tintagel and Bude (Sheet 322 n.s.), The Geology of the Coast between. Freshney, E. C., McKeown, M. C., & Williams, M. 1972.

Torquay (Sheet 350), The Geology of the Country around. Ussher, W. A. E. 1903.

Special Reports on the Mineral Resources of Great Britain

I. Tungsten and Manganese Ores. 3rd ed. 1923 (H. Dewey, H. G. Dines, *et al.*)

II. Barytes and Witherite. 3rd ed. 1922 (G. V. Wilson, T. Eastwood, *et al.*)

IV. Fluorspar. 2nd ed. 1917 (R. G. Carruthers, *et al.*) 3rd ed. 1922 (R. G. Carruthers, *et al.*) 4th ed. 1952 (K. C. Dunham, *et al.*)

IX. Iron Ores (contd). Sundry unbedded ores of . . . and Somerset, Devon, and Cornwall. 1919 (T. C. Cantrill, *et al.*)

XV. Arsenic and Antimony Ores. 1920 (H. Dewey, *et al.*)

XXI. Lead, Silver-lead, and Zinc Ores of Cornwall, Devon, and Somerset. 1921. (H. Dewey)

XXVII. Copper Ores of Cornwall and Devon. 1923 (H. Dewey) (repr. Mining Facsimiles, Sheffield, 1985).

Index

Note: Only the more significant personal names are entered; references to the literature are excluded. Names of minerals, mines and mine localities are indexed in full. The mines are listed together under 'mines (Cornwall),' 'mines (Devon)' and 'mines (not in Cornwall or Devon'. For 'Wheal Abraham' see 'Abraham, Wheal' etc. 'Museums' and 'quarries' are also listed together under their respective heads. Page numbers in italic type refer to illustrations, page numbers in bold type refer to maps, though these characters do not preclude further references on the same page; inclusive page numbers followed by *passim* indicate scattered references.

BACK COVER ILLUSTRATION

Liroconite – group containing the largest known crystal, now at the County Museum, Truro. The Rashleigh ms. catalogue entry reads: '1114 Transparent crystals of Copper Ore in double four sided Pyramids of bright blue colour the largest crystal $\frac{9}{10}$ of an inch on one edge from Wheal Gorland r.r.r.' [r.r.r. = very rare]